OECD Review of Fisheries: Policies and Summary Statistics 2017

OECD
BETTER POLICIES FOR BETTER LIVES

Please cite this publication as:
OECD (2017), *OECD Review of Fisheries: Policies and Summary Statistics 2017*, OECD Publishing, Paris.
http://dx.doi.org/10.1787/rev_fish_stat_en-2017-en

ISBN 978-92-64-28225-4 (print)
ISBN 978-92-64-28226-1 (PDF)
ISBN 978-92-64-28725-9 (epub)

The statistical data for Israel are supplied by and under the responsibility of the relevant Israeli authorities. The use of such data by the OECD is without prejudice to the status of the Golan Heights, East Jerusalem and Israeli settlements in the West Bank under the terms of international law.

Photo credits: Cover © Shutterstock/Cienpies Design.

Corrigenda to OECD publications may be found on line at: *www.oecd.org/about/publishing/corrigenda.htm*.

Foreword

This edition of the *OECD Review of Fisheries* is the latest in a longstanding series of OECD reports covering developments in both production and policies in fisheries and aquaculture. This version broadly covers the period 2015-16 and includes 35 countries and economies, comprising 28 OECD countries as well as a regional chapter covering developments in the European Union. Also participating in this edition are Argentina, the People's Republic of China, Colombia, Indonesia, Lithuania, Chinese Taipei, and Thailand. Together, the participants in this *Review* represent nearly half of global fisheries production, and the majority of production of aquaculture.

This edition follows a new structure and contains much in the way of new content. An entirely new layout of country presentations makes it easy for the reader to find the main policy developments and put their hands on key facts and figures. The presentation of information is improved, with chapter headings that point out key developments and a new table identifying important management features for major stocks. Also new is more evidence on the degree of success that participating countries and economies have had in achieving their quantitative targets for fisheries stock management.

Chapters 1, 2 and 3 of this report comprise the General Survey of fisheries and aquaculture policies. It gives a thematic summary of the information present in the individual chapters for each country or economy. For example, the reader can easily find the latest developments in marine protected areas for each participating economy, or changes in aquaculture regulation in the Country Snapshots (Chapter 4).

A completely revised and expanded presentation of the OECD Fisheries Support Estimate (FSE) database is also new for this year. The FSE is an inventory of policies supporting fishers and the fishing sector that shows the different ways in which budgetary policies are designed and delivered. This publicly available database is the only comprehensive and authoritative source of recent information on these kinds of polices and is a useful resource for policy-makers and academics. More information is available at http://www.oecd.org/agriculture/fisheries/fse.htm.

The print version of this report contains the General Survey of fisheries and aquaculture plus a short summary for each participating country or economy. More in-depth chapters as well as a statistical annex containing detailed background tables with indicators of fisheries support and values for aquaculture production are available in electronic form on the OECD iLibrary (http://dx.doi.org/10.1787/rev_fish_stat_en-2017-en).

The General Survey of fisheries and aquaculture was declassified by the OECD Fisheries Committee at its 120th Session on 14-16 November 2017 and prepared for publication by the OECD Secretariat. Individual Country Profiles were produced in co-operation with the participating country or economy.

Acknowledgements

This *Review* was authored by the staff of the Fisheries Unit of the OECD Natural Resource Policy Division (Roger Martini, James Innes, Woojin Nam, Barbara Hutniczak, Claire Delpeuch and Antonia Leroy) based on contributions provided by participating countries and economies. All statistical work was carried out by Fabiana Cerasa. The project was led by Roger Martini under the overall supervision of Franck Jésus, Head of the Natural Resource Policy Division. Woojin Nam co-ordinated the project. The authors would like to thank Rachel Bae, Ronald Steenblik and the delegates of the OECD Fisheries Committee (COFI) for their comments on previous versions of this work, in addition to Stefania Vannuccini of the FAO for her feedback and assistance with FAO data. The authors would also like to thank the staff of the OECD Public Affairs and Communications Directorate for their editorial guidance and support in the production of the report. The report also benefitted from the substantive and editorial guidance of Natasha Cline-Thomas whose indefatigable support was indispensable for the completion of the project.

Table of contents

Figures

Boxes

Acronyms and abbreviations

AIS	Automatic identification system
CPUE	Catch per unit effort
CCTV	Closed-circuit television
CBFM	Community-based fisheries management
CRC	Cost recovery charge
EUR	Euro
EC	European Commission
EFF	European Fisheries Fund
EMFF	European Maritime and Fisheries Fund
EU	European Union
CFP	European Union Common Fisheries Policy
EEZ	Exclusive economic zone
FSE	Fisheries support estimate
FAO	Food and Agriculture Organization of the United Nations
GFCM	General Fisheries Commission for the Mediterranean
GSSE	General services support estimate or General support to the sector
GFT	Government financial transfers
GHG	Greenhouse gas
GT	Gross tonnes
ha	hectare
IUU	Illegal, unreported and unregulated fishing
ITQ	Individual transferable quota
IOM	Integrated ocean management
ICES	International Council for the Exploration of the Sea
INTERPOL	International Criminal Police Organization
IFFO	International Fishmeal and Fish Oil Organisation
kg	kilogramme
km	kilometre
MCS	Management, control and surveillance or Management, control and enforcement
MPA	Marine protected area
MEY	Maximum economic yield
MSY	Maximum sustainable yield is the largest long-term average catch or yield that can be taken from a stock or stock complex under prevailing ecological and environmental conditions.
COFI	OECD Fisheries Committee
RFMO	Regional fisheries management organisation
STECF	Scientific, Technical and Economic Committee for Fisheries
SDG	Sustainable Development Goal
B_{MSY}	The level of biomass (B) that will produce the maximum sustainable yield (MSY).
B_{pa}	The precautionary reference point for spawning stock biomass (SSB).
F_{MSY}	The rate of fishing mortality (F) consistent with achieving maximum sustainable yield.
SSB	Spawning stock biomass or the total weight of all mature fish in the stock.

TAC	Total allowable catch
TIF	Transfers to individual fishers
UNCLOS	United Nations Convention on the Law of the Sea
UNODC	United Nations Office on Drugs and Crime
USD	United States dollar
VMS	Vessel monitoring system
WCO	World Customs Organization
WTO	World Trade Organization
ZCH	Zone de conservation halieutique

Executive summary

Countries are advancing reforms in their fisheries management systems to improve the profitability and sustainability of their fisheries. They are also working actively to promote the development of aquaculture, which is seen as the future of fish production. These efforts include regulatory improvements and increased spending on research, as well as cost-sharing with the private sector to encourage investment in the sector.

Global aquaculture production already exceeds the volume of catch from wild fisheries, if aquatic plants are included. Average annual aquaculture growth in OECD countries has accelerated and now averages 2.1% per year. Globally, it is even more rapid, at 6% per year. Moreover, average prices of aquaculture products are also increasing, providing an extra boost to the value of the aquaculture sector.

In the case of fisheries, the situation is mixed. Concerns regarding the sustainability of global fisheries are reflected in the fact that some 40% of fish stocks with measurable management targets set by governments are failing to meet those targets. And that story is incomplete, as many more stocks are not measured in a manner that will lead to an accurate assessment.

Harvests of wild fish continue to trend downward in aggregate in OECD countries. In some cases, this is due to the effects of declining fish stocks, while in others greater control of overfishing has also contributed. The increasing value of fish products has moderated the effects of reduced landings on revenue. However, unit values and value of landings have both been declining since their peak in 2011. Overcapacity of fishing fleets is not as serious a problem as it has been in the past, though governments are still investing significantly to reduce the number of fishing vessels in operation.

The Fisheries Support Estimate (FSE) database now inventories budgetary support to fisheries that totals USD 13 billion (EUR 11.7 billion) in 33 countries and economies in 2015. The recent inclusion of the People's Republic of China (hereafter "China") in the database reveals the scale of policies in this important fishing nation—nearly 88% of all support transferred to individual fishers recorded in the database originates in China. In a positive development, China has announced plans for a progressive reduction in their fuel subsidy for fishing vessels.

For most other countries and economies in the database, support to general services to the sector, rather than transfers to individual fishers, dominate. Governments invest a significant amount of resources to this kind of support, which includes management, enforcement, research, infrastructure and marketing. On average, these expenditures by governments equal 16% of the value of landings; that is, USD 1 in every 6 earned by the sector.

While some governments recover these costs from fishers, requiring that the sector support itself, this approach of cost recovery is not commonly applied and accounts for only a small percentage of the total outlay on general services to the sector.

Chapter 1. Production trends in fisheries and aquaculture

This chapter provides an overview of recent trends in fisheries and aquaculture production. The continuing importance of aquaculture's contribution to total production is clear and the average rate of growth in aquaculture output has been 2.1% per year since 2011 for the OECD region. The value of OECD level aquaculture output has grown even faster, averaging 6% per year since 2006, driven by price increases of 4% per year as producers have focused on higher value species. Capture fisheries landings at the OECD level have continued following the long-observed trend of decline and are now at the lowest level observed since 1995. This is the result of both declining stocks and more restrictive fishing policies aimed at ensuring sustainable exploitation. The value of capture landings, which was historically supported by price increases as the quantity of landings fell, has also started to decline more recently as prices have also started falling. Following a long period of decline, recent evidence seems to indicate that the number of fishing vessels has been stabilising at the OECD level in recent years.

The statistical data for Israel are supplied by and under the responsibility of the relevant Israeli authorities. The use of such data by the OECD is without prejudice to the status of the Golan Heights, East Jerusalem and Israeli settlements in the West Bank under the terms of international law.

The value of fisheries landing is down from its 2011 peak

The fisheries reported for this *Review* represent approximately half of the global fisheries in terms of volume of landings. Collectively they have reported landings of 45 million tonnes of fish and seafood in the form of 1 092 different species from all parts of the globe, worth USD 74 billion.

Production of wild-caught fish in OECD countries is considerably below its peak in the late 1980s and continues to decline. Price movements, especially after the financial crisis of the mid-2000s, have acted to mitigate the economic impact of lower landings on the sector, and landing values have been essentially flat since 2006, but with a notable spike in values in 2011 (Figure 1.1).

Figure 1.1. OECD landings in value and volume, 1995-2015

USD millions and thousand tonnes (left axis), USD/tonne (right axis)

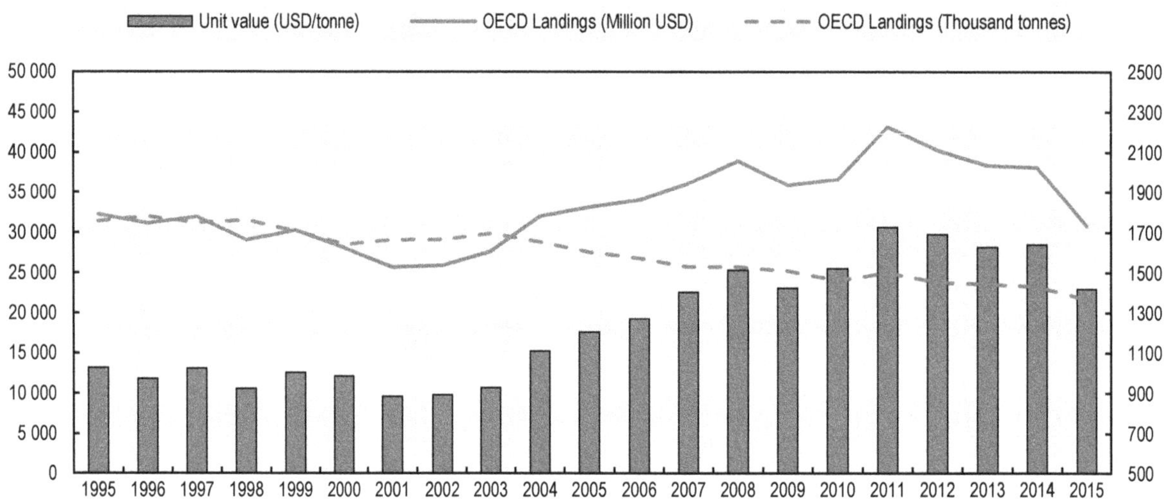

Note: Landings refers to the weight or value of the product at the time of landing, regardless of the state in which is landed (i.e. the fish may be whole, or gutted or filleted). Total OECD landings are constructed from the sum of all reported landings by OECD member countries participating in the Review. This is currently 35 countries and excludes mainly countries with no substantial marine fisheries (such as landlocked countries).
Source: OECD Fisheries Database.

StatLink http://dx.doi.org/10.1787/888933627751

The current situation with respect to landings is the result of both declining stocks and more restrictive fishing policies aimed at ensuring more sustainable exploitation of fish stocks for the long term. While production continues to decline, there is some evidence that the number of fish stocks considered to be overfished is declining, as is the number of stocks where overfishing is taking place.

The countries and economies participating in this review collectively report 1 182 stocks with established targets or thresholds recently subject to quantitative assessment. The majority of these, 725 or 61%, have been assessed as meeting the stated management objectives. Of the stocks deemed to be meeting their objectives, which can vary across countries and stocks, 45% of the determinations have been made on the combined basis of both biomass levels (B) and fishing mortality (F). The remaining determinations are reported to be based upon estimates associated with only one of these indicators, either biomass levels (54%) or fishing mortality (1%).

New Zealand reports stocks at target, below the precautionary and threshold limits as well as subject to overfishing, as a % of all stocks as well as by value and volume (The Status of New Zealand's Fisheries 2016). In terms of the numbers of stocks of known status, by the end of 2016 82% were below the 'overfishing threshold', and 71% were above their management targets. In terms of landings, 94% of assessed landings were made up of stocks that were considered to be meeting their management targets.

The **United States** reports stocks that are overfished and subject to overfishing as a share of all stocks in an annual report to congress that is mandated under the Magnuson-Stevens Fishery Conservation and Management Act (NOAA, 2017[1]). In 2016, 474 stocks and stock complexes managed by federal fisheries management plans were tracked. Of the 235 stocks with known stock status, 197 (84%) were not overfished by the end of 2016 and of the 316 stocks where fishing mortality was known 286 were not subject to overfishing. Of the 339 stocks for which at least an assessment of either F or B was available 167 (49%) were neither overfished nor subject to overfishing and 51 (15%) were either overfished, subject to overfishing, or both.

According to the EU Communication on the State of Play of the CFP and Consultation on Fishing opportunities for 2018, in 2015, 39 of 66 stocks in the North-East Atlantic were exploited within F_{MSY} (59%, up from 52% in the previous year). Median fishing mortality has stabilised at around 1 compared to F_{MSY} (down from 1.58 in 2003 and 1.29 in 2008). The percentage of stocks within safe biological limits is also increasing in the North-East Atlantic. In 2015, 68% of stocks were classified as being within safe biological limits (53% in 2014). The number of total allowable catches (TACs) set in line with maximum sustainable yield (MSY) advice increased to 44, representing 61% of all expected catches in that region, for 20% of the expected catches, there was not MSY advice. The average stock biomass in the North-East Atlantic increased by 35% between 2003 and 2015. In the Mediterranean, fishing mortality remains at levels two to three times higher than F_{MSY} and despite recent improvements, the status of many stocks remains unknown. Furthermore, average biomass declined by 20% between 2003 and 2014.

In the **European Union**, the Common Fisheries Policy (CFP) management objective for fish stocks is for spawning stock biomass (SSB) to attain or exceed the level required to achieve MSY ($B>B_{MSY}$, or its proxy) and for fishing mortality (F) to be at or below that required to achieve MSY ($F<F_{MSY}$).[1] The Scientific, Technical and Economic Committee for Fisheries (STECF) provides stock-level assessment information based on a combination of STECF, ICES and GFCM data (STECF-17-04). This covers 106 stocks from both the North-East Atlantic and the Mediterranean areas, the majority of assessments are with respect to 2015 and the remainder for 2014 and 2013. Of these, F is less than F_{MSY} for 45 of the 106 stocks and above for 61. For B, 26 stocks have $B>B_{pa}$ (within the safe biological limit), 14 are outside this limit and for 66 there is insufficient scientific information to determine the stock biomass size. As regards stocks within/outside CFP requirements, the STECF concludes that the required information is available for 41 stocks; of these 14 stocks are inside CFP requirements and 27 are outside.

Japan reports stocks as being at either "low", "moderate", or "high" levels, (MAFF, 2016[2]). By this method of reporting, out of 84 stocks, 41 were assessed as having a low status, 29 moderate and 14 high in 2016. The stock trend is also reported on, and of the same 84 stocks 16 are increasing, 38 are stable and 30 are decreasing. When this information is combined, 35 of the 84 stocks are at a medium or high level and are either stable or increasing in trend.

Canada reports on both the status of its major fish stocks and on levels of harvest (Environment and Climate Change Canada, 2017[3]) (Environment and Climate Change Canada, 2017[4]; Fisheries And Oceans Canada, 2016[5]). With respect to status, of the 159 stocks that had their status assessed in 2015, 78 were considered healthy, 31 cautious, 19 critical, and for 31 the status was uncertain. With respect to level of harvest, 152 were considered to be harvested at sustainable levels and 7 stocks (4%) were harvested above approved levels. A total of 76 stocks were assessed as having both healthy status and a level of harvest that was at or below an approved level, whilst 52 stocks were known to be either not healthy (cautious or critical), harvested above the removal reference point, or both.

Iceland, through its Marine Research Institute, reports annually on fish stocks (Hafrannsóknastofnun, 2016[6]). The latest report discusses 36 fish stocks, 7 of which are managed on the basis of harvest control rules using quantitative targets and thresholds, and all of these are deemed to be in satisfactory condition.

Norway reported against the status of 75 stocks or stock complexes in 2016, 47 were considered to have been in good condition with respect to both stock status and fishing mortality, 8 to have had low fishing mortality but poor stock status, and 8 to have had both high fishing mortality and a poor stock status. Of the remaining 12 stocks, 5 had low fishing mortality but uncertain status, 1 had high mortality and uncertain status and 6 had both uncertain status and mortality.

Australia has recently started producing a more comprehensive online compendium of stock status assessments (FRDC, 2017[7]), which covers state managed fish stocks in addition to the nationally managed (Commonwealth) stocks that are covered in the Australian fishery status reports. For 2016 it reported against 298 stocks, assessed at the level of biological stocks (85), management units (56), or jurisdiction (34). From this 232 stock status classifications could be assigned and 175 (75%) were reported to have a sustainable status, 26 were transitional-depleting, 9 transitional-recovering, 17 overfished, 5 environmentally limited, 49 undefined. The remaining 13 stocks are considered 'negligible' stocks, as they combined account for only 0.01% of catch. The stocks covered account for approximately 90% of the value of Australia's wild caught fish and of the catch reported in the Status of Australian fish stocks reports 2016, 85% is from sustainable stocks.

Fisheries in some countries specialised on few species, others more diverse

Some countries' fishers pursue a large number of stocks with a broad geographic dispersion, while others are more specialised or localised. This diversity of harvest can be measured using Simpson's diversity index, which measures the probability that two fish drawn at random from the total harvest will be of the same species. This varies from 0 (a very large amount of diversity) to 1 (only a single species is harvested). This index can be seen as a measure of the resilience of the fishing sector in terms of its dependence upon a small number of fish stocks for its revenue.

Using reported landings value data to construct this index, a wide variety of situations in the participating countries or economies is seen (Figure 1.2). In **Argentina,** shrimp hake and squid together account for 75% of total landed value, while prawns represent more than half of all **Estonian** landings. **Canada** lands a good deal of lobster and crab. **Latvia** is also focused on crab and **Lithuania** targets mainly mackerel. On the other end of the scale, **Italy, Japan, United States** and **Korea** harvest a large number of different species, due to the greater diversity of species in their exclusive zones or significant distant water activities.

Figure 1.2. Index of diversity of landings, selected countries or participating economies, 2015

Lower value indicates greater variety of species harvested

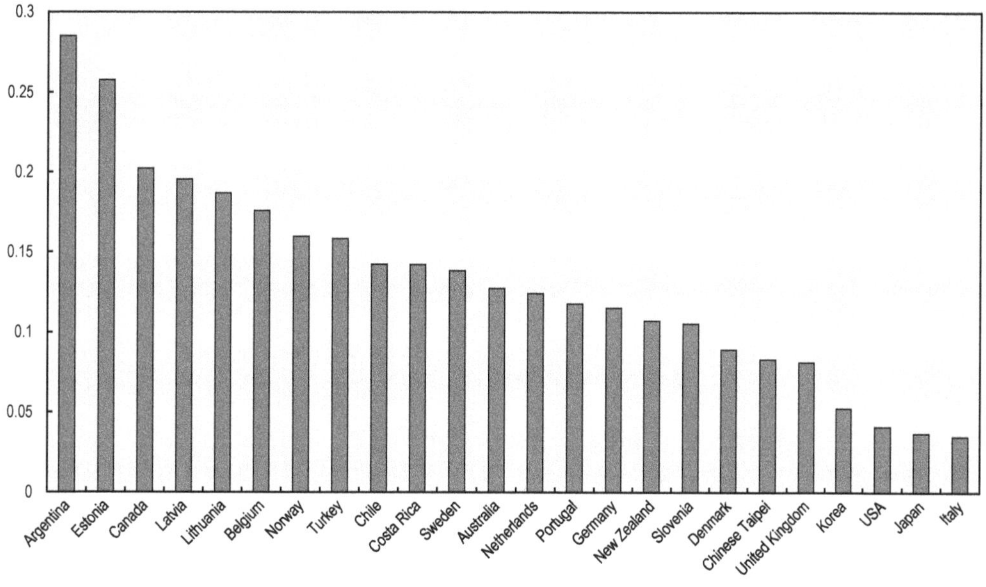

Note: The index is constructed as $\sum\left(\frac{n}{N}\right)^2$ where n is the volume of landings of a particular species and N is the total volume of landings.
Source: OECD Fisheries Database.

StatLink http://dx.doi.org/10.1787/10.1787/888933627770

Aquaculture growth steady, focused on higher value species

Growth in aquaculture production has been famously rapid at the global scale. While it has been more modest on average within the OECD grouping, the rate of growth within this grouping has still averaged at more than 1.5% since 2006 and accelerated to 2.1% since 2011 (Figure 1.3). In value terms performance has been even better; total value has grown by 6% per year on average since 2006 with prices increasing by 4% per year.

While improvements in quality and consumer acceptance have helped improve the price of aquaculture products, at least some of this trend is due to a compositional effect where the production mix is shifting towards higher value species such as salmon and sea bass.

Figure 1.3. OECD aquaculture production in value and volume, 1995-2015

USD billions and thousand tonnes (left axis), USD/tonne (right axis)

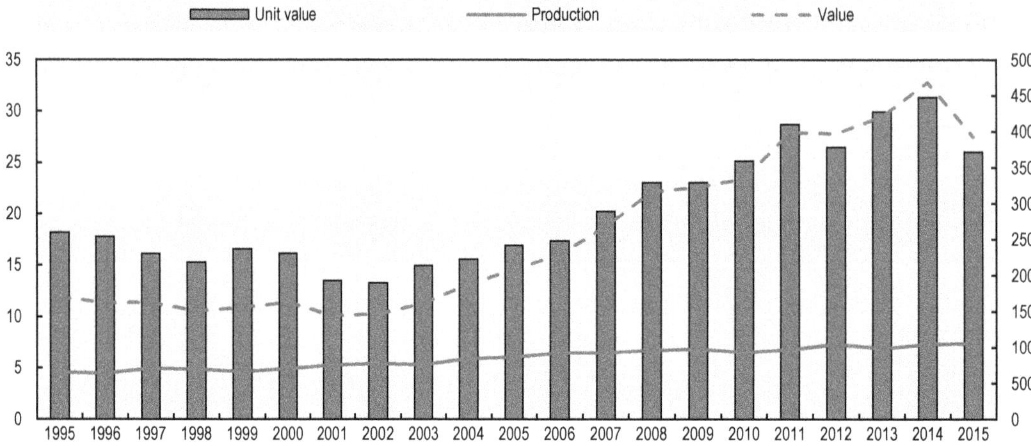

Note: Aquaculture production is the outcome of aquatic organisms' farming. The OECD total is constructed from the sum of all reported landings by OECD member countries reporting landing data in the Review.
Source: OECD Fisheries Database.

StatLink http://dx.doi.org/10.1787/10.1787/888933627789

The People's Republic of China (hereafter "China") dominates global aquaculture production, with a total produced value in 2015 of USD 131.7 billion (EUR 119 billion). This is almost 20 times the value produced in **Chile**, the next highest of those participating in this Review[2]. Among OECD countries, **Chile**, **Norway** and **Japan** are the three largest producers (Figure 1.4).

Figure 1.4. Aquaculture value of production, selected countries and economies, 2015

USD Millions

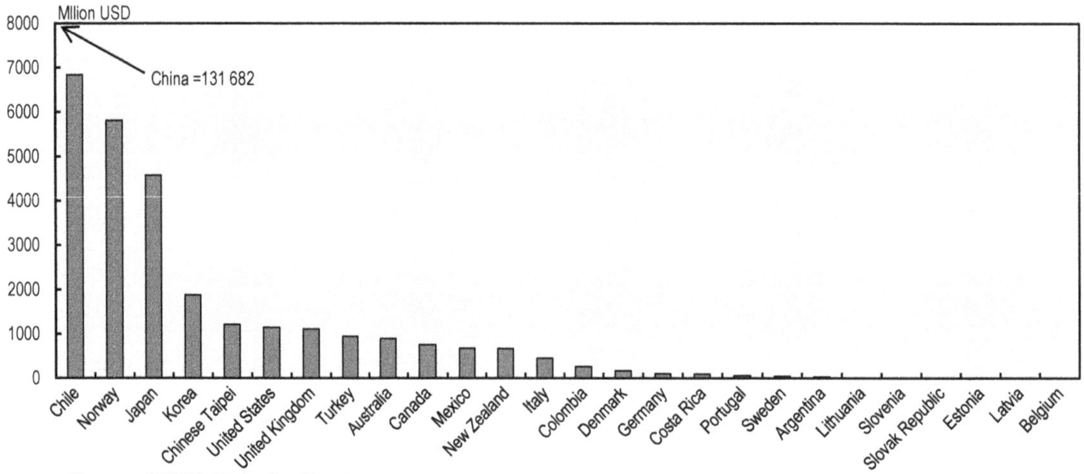

Source: OECD Fisheries Database.

StatLink http://dx.doi.org/10.1787/888933627808

The same index of diversity of production that was presented for fisheries landings can be produced for aquaculture production. This provides an indication of the degree of specialisation of the aquaculture sector of a country as a whole (Figure 1.5). The results show that the three largest OECD producers have very different aquaculture sectors, representing both very specialised **(Norway)** and very diverse **(Japan)** production.

Figure 1.5. Index of diversity of aquaculture production, selected countries or participating economies, 2015

Lower value indicates greater variety of species produced

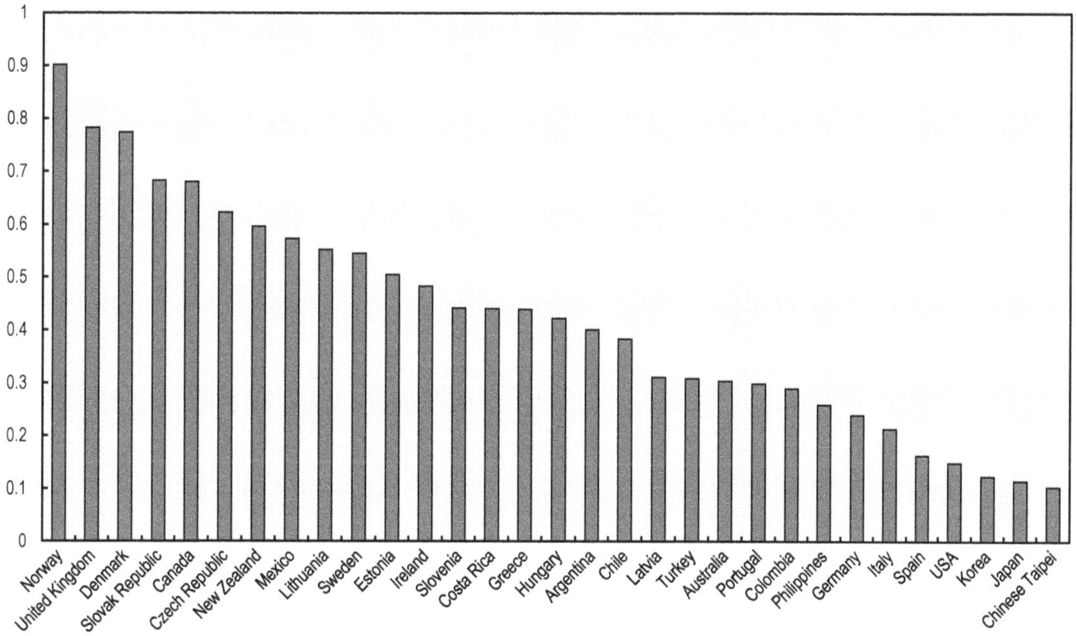

Source: OECD Fisheries Database.

StatLink http://dx.doi.org/10.1787/888933627827

The most highly specialised sectors focus on salmon (**Norway**, **United Kingdom**, **Canada**), rainbow trout (**Denmark**), carp (**Czech Republic**) or a combination of these (**Slovak Republic**). **New Zealand** production is mainly shellfish; mussels, oysters and abalone with some salmon. Those with the most diverse aquaculture are also those who had the most diverse capture sectors, **Chinese Taipei**, **Japan**, **Korea**, **United States**, **Spain** and **Italy**. In **Chinese Taipei** ten species make up 90% of production (out of a total of 53 reported species), led by grouper (*Epinephelus spp*) and oysters (*Crassostrea gigas*). In the case of **Norway**, 95% of the value of production is from Atlantic salmon. The importance of the top ten aquaculture species in **Japan** is similar to Chinese Taipei, but with a smaller total number of species produced (26 reported in 2015).

The pace of decline in fleet sizes is moderating

The size of the fishing fleet has been in general decline for some time as fleets adjust to available fishing opportunities. Of the 25 countries or economies reporting a time series long enough to make the calculation, 22 or 88% have shown a decline in the number of vessels since 2011. This ratio increases to 92% for the time period between 2006 and 2015. The average change in number of vessels for countries or economies reporting data is -2% per year since 2006.

There is some evidence that the rate of decline in the number of fishing vessels is moderating (Figure 1.6). The average rate of decline in number of vessels has moderated since the mid-2000s and is approaching stability in 2013-15, with an average decline in numbers of 0.05% in that period.

Figure 1.6. Average change in number of fishing vessels, 1996-2015

Annual percent change from previous year, average of reporting countries or economies

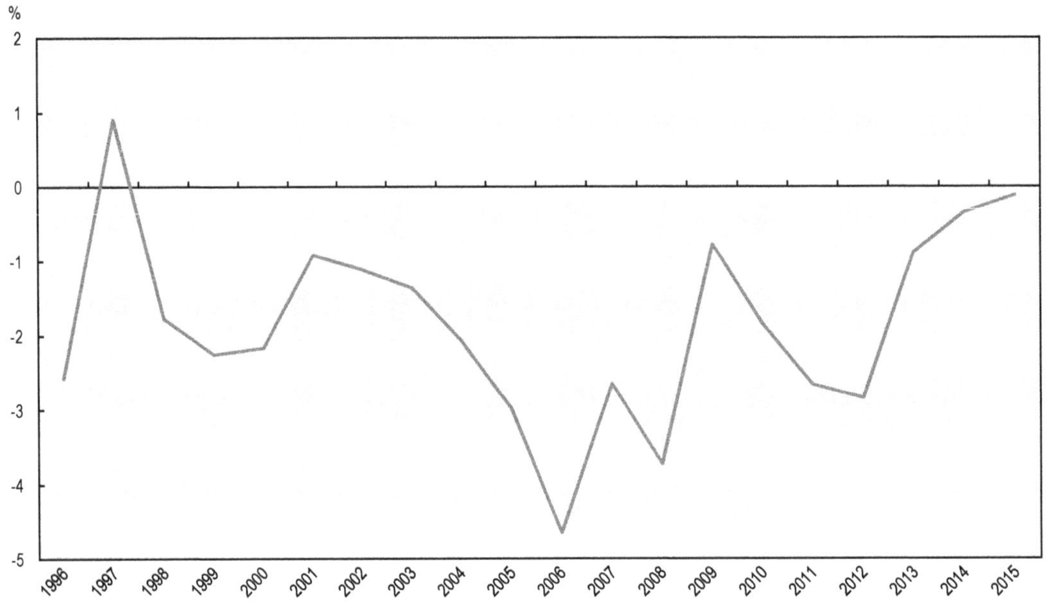

Note: Not all participants have time series of equivalent length, so the average is not composed of the same number of countries or economies in different years. The reported number is the average of the change in each country, not the change in the total fleet.
Source: OECD Fisheries Database.

StatLink http://dx.doi.org/10.1787/888933627846

Notes

1 Article 2 the CFP Regulation (EU 1380/2013)

2 This reported value of aquaculture production is substantially higher than that reported in the FAO. Value data for China can be difficult to estimate accurately as domestic prices of locally-consumed products are imperfectly known.

References

Environment and Climate Change Canada (2017), *Canadian Environmental Sustainability Indicators: Status of Major Fish Stocks*, http://www.ec.gc.ca/indicateurs-indicators/default.asp?lang=en&n=1BCD421B-1. [3]

Environment and Climate Change Canada (2017), *Canadian Environmental Sustainability Indicators: Sustainable Fish Harvest*, http://www.ec.gc.ca/indicateurs-indicators/default.asp?lang=en&n=893AB9F4-1. [4]

Fisheries And Oceans Canada (2016), *Sustainability Survey for Fisheries*, http://, http://www.dfo-mpo.gc.ca/reports-rapports/regs/sff-cpd/survey-sondage/index-en.html. [5]

FRDC (2017), *Status of Australian Fish Stocks Reports*, Fisheries Research and Development Corporation, http://fish.gov.au/ReportStock?kw=&page=1&sort=LatestFirst. [7]

Hafrannsóknastofnun (2016), *State of Marine Stocks in Icelandic Waters 2015/2016 and Prospects for the Quota Year 2016/2017*, Hafrannsóknastofnun (Marine Research Institute), http://www.hafro.is/Astand/2016/fjolrit_185.pdf. [6]

MAFF (2016), *FY 2015 Trends in Fisheries*, Ministry of Agriculture, Forestry and Fisheries, http://www.maff.go.jp/e/data/publish/attach/pdf/index-49.pdf. [2]

Martini, R. (2012), "Fuel Tax Concessions in the Fisheries Sector", *OECD Food, Agriculture and Fisheries Papers*, No. 56, OECD Publishing, Paris, http://dx.doi.org/10.1787/5k9bdccqft30-en. [9]

NOAA (2017), *Status of Fish Stocks 2016*, NOAA Fisheries, http://www.nmfs.noaa.gov/sfa/fisheries_eco/status_of_fisheries/archive/2016/status-of-stocks-2016-web.pdf. [1]

OECD (2006), *Financial Support to Fisheries: Implications for Sustainable Development*, OECD Publishing, Paris, http://dx.doi.org/10.1787/9789264036642-en. [10]

OECD (2017), *Marine Protected Areas: Economics, Management and Effective Policy Mixes*, OECD Publishing, Paris, http://dx.doi.org/10.1787/9789264276208-en. [8]

Chapter 2. Policy developments in fisheries and aquaculture

This chapter provides an overview of the latest major policy developments for countries covered by the Review. Countries are advancing reforms in their fisheries management systems to improve both the profitability and sustainability of their fisheries sectors. In the pursuit of these objectives, and on the basis of their experience to-date, a number of countries are currently reviewing and revising the way in which fishing rights and quotas are both allocated and administered. Countries are also working actively to promote the sustainable development of aquaculture, which is seen as the primary source of future growth in fish production. These efforts include regulatory improvements and increased spending on research, as well as cost-sharing with the private sector to encourage investment in the sector.

The statistical data for Israel are supplied by and under the responsibility of the relevant Israeli authorities. The use of such data by the OECD is without prejudice to the status of the Golan Heights, East Jerusalem and Israeli settlements in the West Bank under the terms of international law.

National plans for fisheries updated in some countries

Many countries maintain national planning documents outlining their main objectives and approaches for fisheries policies. In the European Union (EU), such planning documents are obligatory under the requirements of the Common Fisheries Policy (CFP) and EU members have been renewing theirs as part of the reformed CFP in force since 2013. A Multiannual Plan for the Baltic was adopted in 2016. Further plans for the North Sea, the Atlantic and the Mediterranean were at different stages of development and adoption by December 2016.

The **Australian** Government is currently updating both the Commonwealth Fisheries Harvest Strategy Policy and Guidelines and the Commonwealth Policy on Fisheries Bycatch. Important improvements will include operational guidelines and measures in the Bycatch Policy for assessing its effectiveness over time and providing explicit guidance on how to manage by-product species in the revised Harvest Strategy Policy.

Sweden has initiated development of management targets for nationally-managed stocks in territorial waters, which will be implemented in the coming years. Fisheries management will be linked with targets of environmental policies, particularly through adding management measures to the national programme of action for the EU Marine Strategy Framework Directive.

In 2016, **Korea** adopted a five year master plan (2016-2020) for the development of fishery-related industries and fishing villages. The plan supports sustainability of fisheries, food safety and fishing communities. This new integrated plan aims to increase policy effectiveness— formerly Korea's plan on fisheries consisted of disparate plans.

The National Development Plan 2013-2018 and Sectorial Program of Farming and Fishing of the Secretariat of Agriculture, Livestock, Rural Development, Fisheries and Food (SAGARPA) are the main national guidance documents in **Mexico**. These emphasise the need to improve competitiveness through working towards natural resource sustainability, updating the regulatory framework, integrating production chains, supporting innovative projects through the country and encouraging regional development by promoting small-scale projects within the rural sector.

Some important changes in rights and quota allocation methods

Denmark launched a new scheme for coastal fisheries in 2017 that reallocated extra quotas to the coastal fishery. These extra quotas were transferred from the part of the quota reserved for vessels with individual transferable quotas (ITQs) to strengthen the coastal fishery. Under the new system, there are additional quotas added for cod, sole and plaice. Moreover, for the first time, Norway lobster, turbot, sprat, herring and saithe will be introduced to the scheme.

A new Fisheries Law in **Lithuania** extends the use of transferable rights to fishing quota, already in place for inland waters and the large-scale Baltic Sea fleet, to the long-distance fishing fleet and small-scale Baltic Sea fisheries—essentially covering all fisheries and fleet segments.

A reform of the existing quota management system for capture fisheries in **Norway** is planned to reduce the complexity within the system. A tax to recover resource rents is also under consideration.

A roadmap is being developed in **Indonesia** to introduce quotas, capacity rules, closed seasons, zoning laws and establish multi-stakeholder institutions managing 11 fisheries management areas. New laws are being prepared that will allow the use of rights or privileges in fisheries, an important step towards controlling access to resources.

The challenge of implementing the new landing obligation under the EU Common Fisheries Policy (CFP) spurred the introduction of a new management system for demersal fisheries in **Sweden** that allows limited, short-term transferability of fishing rights. This addresses species with small quotas that are often caught as bycatch with other species. These bycatches may fill up specific quotas and thus become "choke species" that act as a bottleneck and prevent efficient utilisation of quotas. The new regulatory scheme favours small-scale coastal fishers and promotes selective fishing.

Regulatory reforms to make rules more sustainable, efficient and effective

In 2015, **New Zealand** announced a review of the fisheries management system. The purpose of the fisheries review is to ensure the fisheries management system is future focused, and able to provide sustainable fisheries resources for all. A public consultation phase of the review was recently completed and policy development is underway.

In June 2015, the **Australian** Government released the Northern Australia White Paper which sets regulatory reduction goals for Australian fisheries management, promotes greater regulatory co-ordination on aquaculture and seeks greater sharing of fisheries management services to minimise costs through standardised services. The White Paper also sought a Productivity Commission Inquiry to identify duplicative or poorly implemented regulations that were harming investment in Australia's fisheries and aquaculture industries. The Commission's report was handed to the government in 2016.

In 2016, **Canada** began a review of the 2012 Fisheries Act amendments as one of several environmental legislative and regulatory reviews concurrently underway. The amendments established the fisheries protection provisions, which focus the Act's regulatory regime for the protection of fish and fish habitats on managing threats to the sustainability and on-going productivity of commercial, recreational and Aboriginal fisheries.

Italy introduced in 2016 a number of measures intending to enhance small-scale fisheries, including regulations to facilitate the development of consortia allowing small-scale fishers to build self-regulatory frameworks addressing local characteristics of fishing activities. Those are intended particularly for areas that are subject to special protection schemes, as in the case of marine protected areas (MPAs).

Regulations have been simplified in **France** through, *inter alia*, a harmonisation of the designation of competent police and fisheries control authorities and the creation of a single fisheries management advisory board *(Commission consultative de gestion des ressources halieutiques*, or CCRGH) by merging the former commissions responsible for quota monitoring and allocating fishing rights. In addition, in 2016 the procedures for entering and leaving the fishing fleet were reformed to improve the evaluation of the suitability of the application, reinforce consultation of professional committees and competent bodies in the process, and require applicants to justify the granting of a new permit upon the lapse of the existing one.

In 2016, the voluntary vessel buyback programme in **Korea** was revised and became obligatory for certain government designated vessels. The compulsory scheme enhances effectiveness and introduces special vessel types which are likely to cause overfishing and subjects vessels demonstrating a history of illegal, unreported or unregulated (IUU) fishing to the programme as of 2017.

New initiatives in monitoring and control of fishing activity: Progress being made in the fight against IUU

Monitoring and control of fishing activity are actions at port or at sea to observe fishers as they catch, land and market fish products. The tools used to do this are being upgraded in many instances to take advantage of new technologies and methods in order to ensure that fisheries management is fair and effective, as well as to deter IUU fishing.

Denmark has had a risk-based control system for commercial fisheries since 2006. Since 2014 the risk assessment of the fishery and landings of Danish fishing vessels in Danish ports and abroad is fully automated. The Danish AgriFish Agency provides daily updates to a series of risk lists that identify critical vessels. The vessels are given a score between zero and ten, depending on their ability to comply with the rules on a number of selected issues. Fisheries inspectors, receive up-to-date information on PC and smartphones and target the control based on the resulting risk assessment. Denmark has also conducted trials for catch quota registration using CCTV cameras to deter misreporting.

Indonesia, having previously banned large (mainly foreign) vessels from operating in its waters, re-authorised in 2016 fishing vessels above 30 GT, though under more restrictive conditions than had been in place previously. All transhipment at sea is prohibited, and all boats above 30 GT have to install and activate vessel monitoring systems (VMS), of which data is shared with Global Fishing Watch. Vessel registration, the licence system and catch reporting documentation are being revised, eventually allowing domestically-built vessels of up to 200 GT to operate.

A new traceability system came into force in **Lithuania** in 2017. The new system is based on electronic record-keeping of catches and landings and is implemented by the Fisheries Service under the Ministry of Agriculture.

Ireland is investing heavily to further develop capacity in control and enforcement, as well as, data collection. Risk-based inspection of fishing activities and co-operation with other Member States of the EU will be further developed to ensure effective fisheries control and inspection in both Ireland and at EU level. This will be complemented with work to improve efficiency in data collection and management, where there is a focus on improving both scientific assessment of stocks and the knowledge on the state of the marine environment at sea.

In 2016, under the Magnuson-Stevens Act, the **United States** released the final rule establishing the Seafood Import Monitoring Program (SIMP). The Program establishes, for imports of certain priority seafood products, the reporting and recordkeeping requirements needed to prevent illegal, unreported and unregulated (IUU) - caught or misrepresented seafood from entering U.S. commerce, thereby providing additional protections for the national economy, global food security and the sustainability of shared ocean resources. Mandatory compliance for most of the species included in the programme is required by 1 January 2018.

A new application for smart devices for fishers is being implemented in **Estonia** to help speed up the transfer of catch data in real time. This allows the administration to carry out cross-checks of catch data, reduce paperwork and increase transparency. A similar application for coastal and inland fishers is forthcoming and will greatly simplify the production and distribution of obligatory reports.

An electronic fish product traceability system for catches and landings is under development in **Latvia**. The existing traceability system is based on paper documents with a unique traceability number. The new electronic system provides more flexibility, convenience and reduces the amount of paper used.

Regulatory change under **New Zealand's** Future of Fisheries work programme includes use of Integrated Electronic Monitoring and Reporting System (IEMRS), which would involve electronic reporting of catch and fishing effort by commercial fishers, monitoring and verification using automated geospatial position reporting and automated on-vessel cameras. It is expected that geospatial position reporting and electronic catch and activity reporting will be introduced on all commercial fishing vessels from 1 October 2017. Video monitoring will be phased in from 1 October 2018.

In **Germany**, the Sea Fisheries Act and related ordinances were amended in 2015 and 2016. This provides for the involvement of the customs agency and the federal police in fisheries monitoring and control and provides the opportunity to impose penalty points on fishing license holders for serious infringements.

In **Chinese Taipei**, the Five-Year Program for strengthening international co-operation on combating IUU fishing began in January 2016. It contains several elements, including a legal framework, an international economic and trade strategy, rules for traceability of fish and fisheries products, new regulations for control and management of coastal and offshore fishing vessels in port and at sea. This is supported by a National Plan of Control and Inspection (NPCI), established in 2015, to conduct MCS measures on domestic fishing vessels and to ensure their compliance with international regulations and domestic management measures to prevent, deter and eliminate IUU fishing activities. Landing declaration requirements and transhipment controls were also strengthened.

The new Fishing Vessel Monitoring System in **Turkey** (BAGİS in Turkish) was completed in 2016, expanding Automatic Identification Systems (AIS) monitoring to vessels over 12 m (from 15 m previously). 1 350 fishing vessels over 12 m that are collectively responsible for 90% of total production from the seas are now monitored continuously.

Argentina began planning in 2015 to implement a system of on-board cameras, which is now being improved to have the majority of vessels covered in 2018. National control at customs has also been expanded to include more detail for some species, as well as a new catch documentation scheme intended to ease certification of Argentinian catches according to international requirements.

A draft law against IUU fishing was produced in **Colombia** aimed at closing regulatory gaps and strengthening judicial and administrative penal procedures. It recognised the country's responsibility as a flag state by ensuring vessels are subject to national regulatory requirements even beyond national jurisdiction and are considered an extension of Colombian territory when fishing.

The OECD hosted a workshop in 2016 on the subject of tax crimes and other crimes in the fisheries sector, with a view to identifying the links between traditional IUU activity and the criminal activity that may be associated with it (Box 2.1). This workshop demonstrated the need for fisheries managers, police, prosecutors and tax administrations to work together if such activity is to be successfully curtailed. It also demonstrated that countries need to work together because of the multi-jurisdictional nature of criminal activities on the seas.

Box 2.1. OECD workshop on combatting tax crimes and other crimes in the fisheries sector

The OECD hosted a workshop in October 2016 with the FAO and the UNODC that brought together fisheries and tax administrations, enforcement agencies, intergovernmental organisations and non-governmental organisations. Together, they identified best practices in combatting tax crimes and other crimes in the fisheries sector:

Strengthen and enforce legislation

- A strong legal framework including fisheries law and relevant regulations, criminal codes, tax legislation, anti-corruption law, labour laws, and organised crime laws is needed. The framework should reflect the state's responsibilities as a flag state, coastal state, market state or port state and for their own nationals.
- Illegal operators exploit gaps in the law at every part of the value chain from fishing boat to consumer. Sector participants should build transparency into their operations to identify where crimes are likely to occur.
- Measures must be tailored to the varied legal powers, procedures and governance capacity of countries.

Build and improve inter-agency co-operation

- Fighting these crimes will require co-operation between police, prosecutors and fisheries managers within and across countries to tackle their increasingly transnational nature.
- An umbrella framework for co-operation can work across borders to combat fisheries related crimes by promoting the exchange of data and information.

Work with and receive support from international organisations

- Partnerships across organisations - such as the OECD, FAO, UNODC, WCO, INTERPOL and RFMOs - can help establish international co-operation and governance frameworks.
- Build on existing measures to fight illegal, unreported and unregulated (IUU) fishing and fisheries related crimes, such as the FAO Agreement on Port State Measures Agreement, the UN Convention against Transnational Organized Crime; and the OECD's work on best policy practices.

Involve stakeholders

- Private sector actors, such as seafood processors or financial service providers, can help ensure that their activities do not facilitate criminal activity.

Increase awareness and understanding

- Increasing awareness of the damage and costs of fisheries-related crimes can motivate a strong response by governments.
- Lessons can be learned from other countries' experiences fighting fisheries and other environmental crimes.
- Improved training and more resources for investigations are needed to build capacity to fight tax crime and other crimes in the fisheries sector.

Source: OECD (2016), Key Findings and Discussion Summary, www.oecd.org/tad/events/Fisheries-crimes-conference-summary.pdf.

Marine protected areas and spatial planning expanding in several places

Progress in expanding the coverage of marine protected areas is underway. With a push from the Sustainable Development Goals (SDGs) their global coverage is expected to increase even further. Marine protected areas are one of the policy instruments available to help ensure the conservation and sustainable use of our vast yet vulnerable ecosystems, in particular to over-fishing and exploitation and habitat destruction. In addition to protecting rare and threatened species and their habitats and other areas of ecological importance, MPAs can help ensure the sustainable provision of multiple other ecosystem services that are fundamental for human well-being, including for fisheries, coastal protection (buffering against storms and erosion), tourism and recreation (OECD, 2017[8]).

During 2015-16 the **Australian** Government undertook a review of Commonwealth Marine Reserves and the report was released by the Director of National Parks in September 2016. The review considered the views of commercial, recreational fishing and other marine users regarding access to marine resources, accounting for desired conservation outcomes. Based on the outcomes of this review, new management plans are being developed and are expected to be finalised during 2017.

On World Oceans Day, 8 June 2016, **Canada** announced its commitment to increase marine conservation from approximately 1% of Canada's marine and coastal areas to 5% by 2017 and 10% by 2020. In October of 2017, Canada announced that it had achieved the 5% target.

In 2016, **Denmark** held consultations with the European Commission (EC), ICES, the regional advisory councils and Member States with regards to an adequate protection of reef structures in the Kattegat and the western Baltic Sea for a total of seven sites. Ten other sites were decided in 2016 to be protected as of 1 January 2017. By the end of 2016, 31 sites were protected. Denmark is also in progress of developing a national strategy for the protection of harbor porpoises in all Danish waters. The strategy will focus on the areas with high risk of porpoise bycatch and it is expected to be implemented by 2020.

Spatial planning in the **Netherlands** is reflected in the 2016-2021 North Sea Policy Document. A ban on fisheries in protected Natura 2000 areas and in the areas that are destined for either offshore wind energy or oil and gas extraction has recently taken effect. Although the restricted area is a small portion of the North Sea, it often covers locations that have been traditionally considered fishing grounds.

Indonesia has committed to establishing over 20 million ha of marine protected areas (MPAs) by 2020. In early 2017, MPAs already covered over 17.97 million ha.

The *Zones de conservation halieutiques* (ZCH) in **France** constitute a new category of protected marine area created in 2016 for the recovery of biodiversity, nature and landscapes. This new tool is dedicated to the protection of the functional fisheries areas essential to the productivity of stocks, such as spawning grounds and nurseries. The classification of an area as a ZCH is carried out by a decree specifying the perimeter of the zone, the duration of the classification, the conservation objectives and the authority in charge of the implementation of the conservation measures. At the beginning of 2017, almost 22% of waters under French jurisdiction were protected, through the various tools for protecting biodiversity, compared with 4% at the beginning of 2014.

The marine protected area programme in **Korea** has expanded and now includes 27 registered areas. A recovery project for tideland ecosystem aims to rehabilitate marine ecosystems with 11 sites completely restored and four more expected in 2017.

Since 2015, owners of trawling licences in **Sweden** are obligated to take a course on the regulations specific for the Koster-Väderöfjorden MPA. Trawling was at the same time banned in five additional areas in the MPA (Kungsviksflaket, Björns, Yttre Vattenholmen, Ullvillarna/Berggylteskär, Grisbådarna) and only vessels equipped with AIS (Automatic Identification System) are permitted to fish for Nordic prawns. In 2016, a no-take zone was introduced within the Bratten MPA area of Skagerrak.

In accordance with its Basic Policy for Fisheries (revised in 2017), **Japan** has been implementing measures for ensuring sustainable fisheries while conserving marine ecosystems, which includes, *inter alia*, conservation of seagrass beds and tidal flats and, use of MPAs to protect spawning areas and to recover fish stocks.

The proposed Marine Protection Act, will replace **New Zealand's** existing framework for creating marine reserves, and include a mix of legislative tools that provide a graduated system of marine protection and allow trade-offs between competing uses. These tools include marine reserves, species-specific sanctuaries, seabed reserves and recreational fishing parks.

Several new policies introduced to improve environmental performance and protection

Latvia is currently working on a bio-economic strategy which will incorporate elements from Blue Growth. This is a complex strategy requiring co-operation between multiple ministries. This makes the process slow and the work is taking longer than anticipated.

Specific measures applicable to the **Portuguese** fishing fleet were adopted with the aim of protecting the seafloor of the continental shelf against any adverse impacts that may be caused by bottom fishing gears. The use of fishing gears that may adversely impact on vulnerable deep sea marine ecosystems is prohibited and information on natural resources of the seafloor will be collected to improve the available scientific data on the marine environment.

Fuel saving is a major goal in the **Netherlands**. Investments in fuel saving gear have led to a 24% decrease in 2014 of fuel consumption (in litres) per fishing day compared to the average between 2008 and 2013. Improvements in fuel economy have been widely observed in the EU fleet: the yearly aggregated fuel consumption of the EU fishing fleet reduced 15% in the period 2008-2015.

A project to remove abandoned fishing gears from the seas in **Turkey** swept 15 000 acres of water area at different locations, removing a total of 270 000 m^2 of ghost nets in different lengths and 1 700 fishing gears such as traps and fyke nets.

In 2016, the **United States** established a rule to complete the implementation of the Marine Mammal Protection Act (MMPA) which imposes ban on imports of fish that result in bycatch of marine mammals in excess of US standards.

The Sustainable Fisheries Framework (SFF) in **Canada** provides the policy basis for ensuring that fisheries are conducted in a manner which supports conservation and sustainable use, while also promoting economic prosperity. The implementation of

SFF policies has created concrete steps for policy-makers to take towards the application of an ecosystem approach to fisheries management.

Argentina is introducing ecosystem based management to its fishery resources. In 2016, a project was initiated with the Global Environment Facility (GEF) called "Governance Strengthening for the Management and Protection of Coastal- Marine Biodiversity in key ecological areas and the implementation of the Ecosystem Approach to Fisheries (EAF)".

Promoting aquaculture production high on the policy agenda

Governments, virtually without exception, have objectives and policies in place to promote the development of aquaculture. This is commonly motivated by its economic potential; aquaculture has been one of the fastest-growing forms of food production for many years and growth in aquaculture can provide jobs and regional development opportunities in areas with few economic alternatives. The emphasis is shifting from stimulating growth to putting the sector on a sustainable footing for the future by addressing environmental limits and focusing on new production technologies to increase competitiveness.

Development of aquaculture is supported by the **European Union** and is part of the reformed Common Fisheries Policy. In 2015, all Member States finalised Multiannual National Strategic Plans for the promotion of sustainable aquaculture in their countries (some specific details are in the following paragraphs). Six European Union Member States (**Czech Republic**, **Germany**, **Lithuania**, **Poland**, **Slovak Republic** and **Slovenia**) allocated the largest share of their European Maritime and Fisheries Fund (EMFF) spending to fostering environmentally sustainable, resource-efficient, innovative, competitive and knowledge-based aquaculture (Union Priority 2). All in all, EUR 1.2 billion (USD 1.3 billion) of EMFF resources are specifically allocated to aquaculture (21% of the funds available).

The National Aquaculture Development Programme (NADP) was established in **Greece** to identify national objectives for the development of the sector. It establishes the National Council for Aquaculture which provides for the formulation and implementation of the NADP and possible institutional reforms, including a "one-stop-shop" principle, with one single licensing authority (the Decentralized Authorities) executing the whole process.

In its 2015 National Strategic Plan for Sustainable Aquaculture Development **Ireland** introduced a set of 'Guiding Principles for the Sustainable Development of Aquaculture', and 'Scale Limits for Individual Offshore Salmon Farms'. A review of Ireland's regulatory framework for aquaculture licensing was initiated in 2016 with recommendations due in the first half of 2017.

The National Strategic Plan for Aquaculture in **Spain** includes among its main objectives improving environmental sustainability. Various actions have been designed to ensure the enhanced compatibility of the aquaculture with the goals of biodiversity conservation. Spain is also in the process of preparing a new aquaculture law. This new legal framework will support progressive development of the sector. Improvements to aquaculture information systems are also planned to unify information currently dispersed among different sources and establish a single register of authorised entities.

The **Danish** government published a "Growth plan for aquaculture" in 2016. Growth and development of the sectors are limited primarily by the possibility of discharging nitrogen. It was decided to increase the allowed nitrogen discharges for both land-based and sea-based farms making increased production possible. The plan also incentivises land-based farmers to convert to the more environmental friendly recirculation systems (less discharge of nitrogen per produced fish volume) to allow expansion of aquaculture production in longer term. Furthermore, it was decided to allow increased nutrient discharge (in relation to what could be allowed within the previous legislation framework) from new sea-based farms, when this is accompanied by compensatory measures (such as mussel farming) that remove a specified part of the increased amount of nutrients.

Australia has developed a National Aquaculture Strategy in consultation with the aquaculture industry and the state and territory governments that is expected to be released in 2017. The aim of the Strategy is to double the value of the aquaculture industry by 2027.

The latest Development Plan (2014-2018) in **Turkey** prioritises development of aquaculture, specifically developing breeding stock management and breeding programmes, encouraging environmentally-friendly aquaculture systems, ensuring continuous supply of feedstuff for aquaculture production and investigating alternative feedstuffs sources.

In 2014, **Colombia** adopted a National Plan for Sustainable Aquaculture (PlaNDAS), which identifies areas that are most suitable for aquaculture development. PlaNDAS identified low economic productivity as a key challenge for aquaculture as it affects competitiveness and reduces profitability, even as the sector has to contend with low-cost imports of fishery products and a small domestic market.

The **United States** released its 5-year Strategic Plan to guide development of sustainable marine aquaculture and established the target of expanding marine aquaculture production at least 50% by the year 2020.

Korea enacted the Fisheries Seed Industry Promotion Act in 2015. Based on the law, five year master plans and annual plans for implementation will be established. To balance demand and supply of fisheries seed, the government can monitor facility capability, output, global market situation and share information to the public. Also, the act introduces the system of fisheries seed promotion such as registering of female brood fish and the broodstock rental and exchange system between public owners and private properties.

Aquaculture regulations are maturing as the sector becomes established

Icelandic authorities have implemented regulations that obligate sea cage companies to apply to the Norwegian NS-9418:2009 standard on all equipment used for sea farming in Iceland to minimise the risk of escapees. All Icelandic sea farms are obliged to contribute to a special aquaculture environmental fund. The objective of the fund is to promote research concerning the sea farming areas and the main beneficiaries of the fund has been The Marine and Freshwater Research Institute.

In July 2015, after four years of effort, the Aquaculture Activities Regulations (AAR) came into effect in **Canada**. These are the first national regulations of their type. The regulations build on existing provincial and federal regimes and allow for better co-ordination of the management of pesticides and drugs used to treat fish for pests and pathogens.

Planning is underway to refresh **New Zealand's** Aquaculture Strategy by early 2018. Part of the process will include an evaluation of the current Aquaculture Strategy and Five-year Action Plan to guide sustainable growth of the aquaculture sector, originally adopted in 2012. New Zealand is also undertaking a public consultation on a National Environmental Standard for Marine Aquaculture (NES: Marine Aquaculture), which would replace existing regional council rules. The proposed NES: Marine Aquaculture seeks to provide a more efficient and certain consent process for managing existing marine farms within environmental limits; and implement a nationally consistent framework for biosecurity management on all marine farms.

Ireland's National Strategic Plan for Sustainable Aquaculture Development includes a review of the regulatory framework for aquaculture licensing and associated administrative procedures. This review was initiated in 2016 with the setting of its terms of reference and the appointment of a 3-person Independent Review Group.

In 2016, the first regional regulatory programme in the **United States** for offshore aquaculture in federal waters was implemented. The Fishery Management Plan establishes a permitting process to manage the development of an environmentally sound aquaculture industry in federal waters of the Gulf of Mexico.

A new system for regulating growth in the salmon farming industry was introduced in **Norway** in 2017. The new system is based on environmental indicators and production areas. It introduces Innovation Licences, a hybrid of research licences and regular commercial licences, in order to stimulate productivity-enhancing innovations.

The **Slovak Republic** as part of their National Plan for Aquaculture is planning to reform the current law regulating the certification of aquaculture farms. The main changes are focused on clarifying the documentation necessary to gain approval in order to ease access to the sector and encourage new entrants.

In 2017, the OECD published the report "Red tape and administrative burden in aquaculture licensing" which looked into the relationship between regulatory systems and sector growth (Box 2.2). While the link between the degree of "red tape" and the rate of growth in production was unclear, it highlighted some of the other benefits that can arise from a systematic approach to developing effective regulations, and that there is no inherent trade-off between the stringency of regulations and their administrative burden.

Box 2.2. Aquaculture licensing

The study used survey data on aquaculture licencing systems to investigate the design of licensing schemes. It builds on existing OECD work on Product Market Regulation (PMR) indicators and the Burden on the Economy of Environmental Policies (BEEP), which provide a ready framework for analysis of the burden of licencing and permitting systems for aquaculture. The results highlight the need for better measurement of environmental outcomes of aquaculture operations to support regulatory evaluation and review.

Key findings

- Experts from governments agree that aquaculture licensing systems are generally achieving their goals for environmental and social protection, although in the majority of cases these systems are not currently as simple as they could be.
- The extent of administrative burden is highly variable, with no corresponding evidence that effectiveness is equally variable. This suggests that there is opportunity to reduce levels of administrative burden on aquaculture producers without sacrificing regulatory quality in the process.
- Administrative burden has some negative correlation to growth; licensing systems displaying high levels of administrative burden tend to have lower rates of growth. Higher levels of burden also imply higher costs for enterprises.
- Quality governance, as measured by institutional oversight, consultation and planning, helps to reduce administrative burden. Additionally, having formal processes to evaluate the effectiveness of regulation and review its impacts support good governance. The survey results show limited use of evaluation and review across countries, suggesting an opportunity for beneficial reforms by increasing their use in order to help reduce administrative burden over time.
- The observed significant negative correlations between the level of administrative burden to entry and both governance quality and evaluation and review are consistent with experiences in other sectors and provides empirical support to the use of these practices as a standard feature of policy design.
- It was not possible to compare the attributes of licensing systems with environmental performance due to a lack of suitable indicators. This highlights the need to measure environmental performance if regulatory effectiveness is to be evaluated further.

Source: Innes, J., R. Martini and A. Leroy (2017), "Red tape and administrative burden in aquaculture licensing", OECD Food, Agriculture and Fisheries Papers, No. 107, OECD Publishing, Paris, www.dx.doi.org/10.1787/7a56bfbc-en.

Aquaculture information systems also under development

In the **Netherlands**, a website has been set up that centralises information on licensing for aquaculture, including all the administrative documents which have to be filled by producers to obtain or renew a license.

Innovation

Australia is using its 2016-20 National Fishing and Aquaculture Research Development and Extension Strategy to plan for how research, development and engineering in fishing can become more focused, efficient and effective. The 2015-25 National Marine Science Plan outlines the research, infrastructure, skills, partnerships and investment needed to drive the changes required over the next ten years and deliver the best possible long-term returns to Australia.

The Primary Growth Partnership (PGP) in **New Zealand** is a joint venture between government and industry, that invests in long-term innovation programmes to increase the market success of the primary industries. An expected change by the end of 2017 under The Future of Fisheries work programme will enable innovative trawl technology (EITT) that focuses on reducing bycatch of undersized fish, reducing the quantity of unwanted fish, and enabling fishers to derive maximum benefit from their catch by improving catch quality.

Support for innovation in aquaculture is an important element of the EMFF, with 24 EU Member States allocating EMFF funds to it.

Restocking exercises are a common feature of stock recovery plans in several Asian countries

Governments can undertake several actions to aid the recovery of depleted fish stocks. Aside from better control of fishing pressure, actions include enhancing the productivity of the ecosystem by restoring its functions and releasing hatchery-grown fish to augment wild populations. Such actions can be linked to advances in aquaculture technologies that make producing juvenile fish for release feasible.

In 2015, **Japan** established the 7th Basic Plan for Fish Farming in Japan, which is revised every five years. The plan includes, among other elements, integrated implementation of releasing juveniles for stock enhancement and capturing fish.

Artificial restocking, artificial reefs and seasonal closures are some common measures adopted for further restoration and enhancement in the **People's Republic of China** (hereafter "China"). New reserves for aqua-germplasm resources are set up every year. Efforts to better measure fisheries stock status and identify spawning grounds continue to improve. Marine ranches (areas with artificial restocking and systematic management) are also being established for resource conservation and environment protection. Twenty demonstration marine ranches have been established in 2015, with more planned in the future based on lessons learned.

A project for generating fishery resources in **Korea** includes marine ranching and marine afforestation. This is in the context of plans to rehabilitate marine ecosystems – 11 sites have already been completely restored with four more expected in 2017.

Chinese Taipei established an activity called "2016 Love the Ocean, Fishery Sustainability: The Fingerling Release Activity." The government works with different universities regarding long-term plans for such restocking activities. "The Coastal Blue Economic Growth Plan" is a project to produce high-value fish such as blackfish (*girella punctate)*, and has been successful in producing good quality fingerlings.

Policy initiatives aimed at improving domestic market balance are tied to objectives for self-sufficiency

While most countries have policies in place promoting aquaculture growth as a means of economic and regional development, these are not usually aimed at changing domestic market conditions. However, several countries have policies designed to achieve a better balance between domestic supply and demand for fisheries products. Some market disruptions have been observed, particularly in the Baltic States, as a result of the loss of the market for fish products in the Russian Federation (Box 2.3).

The national growth objective in the **Slovak Republic** is to achieve 80% freshwater fish farming self-sufficiency in volume by 2020, up from the current level of 40%. The goal is to increase aquaculture production and achieve self-sufficiency in freshwater fish production and enhance competitiveness of the aquaculture sector, in part by introducing new species such as perch, pike, sturgeon and freshwater catfish.

Argentina is promoting local consumption of seafood. A policy is being developed that includes certification scheme for quality products of national origin, labelled "*Alimentos Argentinos*".

Korea promotes the aquaculture industry as a means of maintaining production growth aligned with domestic demand. Part of this effort encourages the development of eco-friendly aquaculture technologies to ensure that growth is sustainable for the long-term. The government also provides monitoring and outlook data for produce and market prices for 14 species to fishers, distribution dealers, and consumers. Fishers can adjust their production based on this system.

Mexico promotes fish consumption and aims to stimulate demand for domestic fishery and aquaculture products. This is done through, *inter alia*, campaigns promoting local seafood, monitoring seafood consumption and organising commercial and gastronomic events.

China maintains considerable policy scope over production, which one aspect of policymaking is to ensure that aquaculture species produced are desired by the market.

To help fishers manage their businesses, **Japan** has established Annual Guidelines for Aquaculture Production for the harvest season, since 2014. Guidelines indicated the supply target quantity on a national scale, based on the Review Meeting on Demand and Supply of Farm Raised Fish. The 2016 guideline proposed 140 000 tonnes for amberjack and 72 000 tonnes for sea bream in supply.

Portugal has the highest per capita fish consumption rate in the EU (54 kg/year in live weight terms) and the overall import to export coverage rate for fish products is 58.4%. Portugal promotes both growth in output and diversification of the types of fish produced in the aquaculture sector to satisfy increasing national demand.

Box 2.3. Russian ban impacting markets

On 7 August 2014 the Russian government imposed an import ban on agricultural and processed food exports from the European Union (EU), the United States, Norway, Canada and Australia. The imposition of the ban was in response to sanctions previously imposed by Western economies on Russian business interests in connection with the crisis in eastern Ukraine. The ban includes most fish products.

The ban has most strongly impacted EU Member States for whom the Russian Federation is a traditional market. The best example may be smoked canned sprats from Latvia where the Russian Federation was the primary market, but other pelagics and farmed salmon have also been affected. Exporters have in some cases seen decreased market prices for their products as alternative destinations such as the Ukraine are more price-sensitive.

One of the intended outcomes of the ban is increased Russian production of agricultural and food products. Russian fisheries production has indeed been increasing, though this has more to do with on-going changes in the Russian fleet and the status of the fish stocks they target.

Recently the Russian government announced an extension of the embargo until the end of 2018.

Chapter 3. The FSE database and indicators of policy support to fisheries

This chapter provides the latest available data on support to capture fisheries. The Fisheries Support Estimate (FSE) database now inventories budgetary support to fisheries that totals USD 13 billion (EUR 11.7 billion) in 33 countries and economies in 2015. For the first time, data for the People's Republic of China (hereafter "China") is included in the database, revealing the scale of policies in this important fishing nation. Nearly 88% of all support transferred to individual fishers recorded in the database originates in China. In a positive development, China has announced plans to progressively reduce this subsidy. For most other countries and economies in the database, support to general services to the sector, rather than transfers to individual fishers, dominate. Governments invest a significant amount of resources to this kind of support, which includes management, enforcement, research, infrastructure and marketing. On average, these expenditures by governments equal 16% of the value of landings; that is, USD 1 in every 6 earned by the sector. While some governments recoup these costs from fishers, this approach is not commonly applied and accounts for only a small percentage of the total outlay on general services to the sector.

The statistical data for Israel are supplied by and under the responsibility of the relevant Israeli authorities. The use of such data by the OECD is without prejudice to the status of the Golan Heights, East Jerusalem and Israeli settlements in the West Bank under the terms of international law.

New countries in the FSE change the picture of global support

In May 2017, the OECD released a first study using the new Fisheries Support Estimate (FSE) database, replacing the old Government Financial Transfers (GFT) classification that was used in the last OECD Review of Fisheries in 2015 (Box 3.1).

At the time that report, which included data from 31 countries, mainly members of the OECD, showed that more than 90% of budgetary support provided by governments was in the form of general services to the sector (GSSE). This kind of support offers benefits to the sector generally but does not make payments directly to fishers. The largest share of such support is for the provision of management services; that is, the cost of administration and operation of the management system.

Programmes that make payments directly to fishers, called transfers to individual fishers (TIF) in the FSE, were only a small share of the total and involved support to help fishers invest (for vessel construction or modernisation), leave the sector (decommissioning schemes) or to support their income (unemployment benefits).

The report did not contain information on fuel tax concessions, a form of support that has been the subject of some attention in recent years. This is because while these programmes have the same basic effect—reducing the cost of fuel purchased by fishers—they vary in their sector specificity. That is, in many countries these are not fisheries policies, but part of a larger scheme. As the FSE collects only programmes specific to fisheries, the coverage of this form of support is highly uneven across countries and so is omitted from reporting. The last time a systematic estimate of the value of fuel tax concessions was made was in 2008 (Martini, 2012[9]). That study found the value of these concessions to be around USD 2 billion (EUR 1.3 billion).

Much has changed with the recent update of the FSE database, which now includes 33 economies, seven of which are outside the OECD. Most significantly, the Peoples Republic of China (hereafter "China"), the largest fishing nation in the world, now appears in the FSE for the first time, changing the global coverage and the picture of support significantly. Costa Rica also enters the database for the first time, joining Colombia, to help improve the understanding of support in less-developed countries with smaller fisheries.

The most striking aspect of this is how it changes the scale of support and its composition. The largest single support policy in the FSE is now the fuel subsidy in China, which delivered USD 6 billion (EUR 4.6 billion) of support in 2014, the last year for which data is available.[1] Support in China in 2015 has been estimated on the basis of government announcements that the level of support will be reduced to 40% of the 2014 amount by 2019.

This one policy accounts for 45% of the total budgetary FSE and changes the share of TIF in the FSE from less than 10% previously to more than 40%. The scale of this one programme is such that its changes drive the trends in the overall FSE. For this reason, some data is reported both with and without China included in order to focus on some of the other policy trends in the FSE.

Support in the FSE now totals almost USD 13 billion

The current version of the FSE database contains data for 2009 to 2015, with some earlier data available for certain programmes. Budgetary policy support to the capture fisheries sector as reported in 26 OECD countries resulted in net transfers totalling a reported USD 7.04 billion (EUR 6.34 billion) in 2015. In some of the discussion below, this group is used to produce an OECD average.

Total budgetary support reported in the FSE database including the seven non-OECD countries currently participating is almost USD 13 billion (EUR 11.7 billion)[2]. The majority of this was provided in the form of support to general services (GSSE) which, after accounting for cost recovery charges (CRC) of USD 170 million (EUR 150 million), are estimated to have totalled USD 7.04 billion (EUR 6.35 billion)[3]. USD 6.09 billion (EUR 5.49 billion) of support was provided in the form of transfers to individual fishers (TIF)[4]. In 2015 88% of all reported TIF was attributable to China's fuel subsidies policy.

Instances of non-reporting have left some gaps, particularly for categories relating to support in the form of management and enforcement expenditures. All fisheries systems face costs associated with management and enforcement, so to ensure that aggregate levels of support reflect this and to avoid having missing data determine the TIF/GSSE ratio in the FSE, these gaps are currently filled with estimates. The missing values for management and enforcement support were estimated by deriving the average ratio of support to value of landings observed in reporting countries and applying this to the value of landings in non-reporting countries. The Secretariat is working with reporting countries to fill these gaps and eliminate the need for estimates[5].

Support as a proportion of the value of landings has decreased overall in OECD countries

The FSE offers a partial view of global support, as not all countries are included in the database. This makes assessing trends over time less conclusive, as the number of countries for which data is available in the FSE changes. In addition, some gaps in reporting in some years may affect trends in the data.

The amount of support provided can vary considerably from year-to-year. This can be the result of the implementation of a multi-year programme, a result of changing economic and environmental conditions or other reasons. Wherever possible the base period when considering individual country data has been set as the annual average of 2009-11 and is compared with the reference period of 2015. As additional data is collected and a longer time series develops, the trends in the data will become clearer and more robust.

The recent inclusion of data relating to China has changed the pattern of budgetary support in aggregate. The underlying pattern of declining overall support and a shift towards GSSE policies still exist, but now does not begin until 2012 and 2014, respectively. This is due to a progressive and substantial increase in TIF by China between 2009 and 2013, which has since been falling.

The data show that, in general and in aggregate, reporting countries spend on support an amount approximately equal to 20% of the value of landings (Figure 3.1). This amount has been declining over time to around 16% in 2015, though adding China flattens this trend considerably.

Figure 3.1. Evolution of budgetary support to fisheries, 2009-15

Percentage of gross value of landings

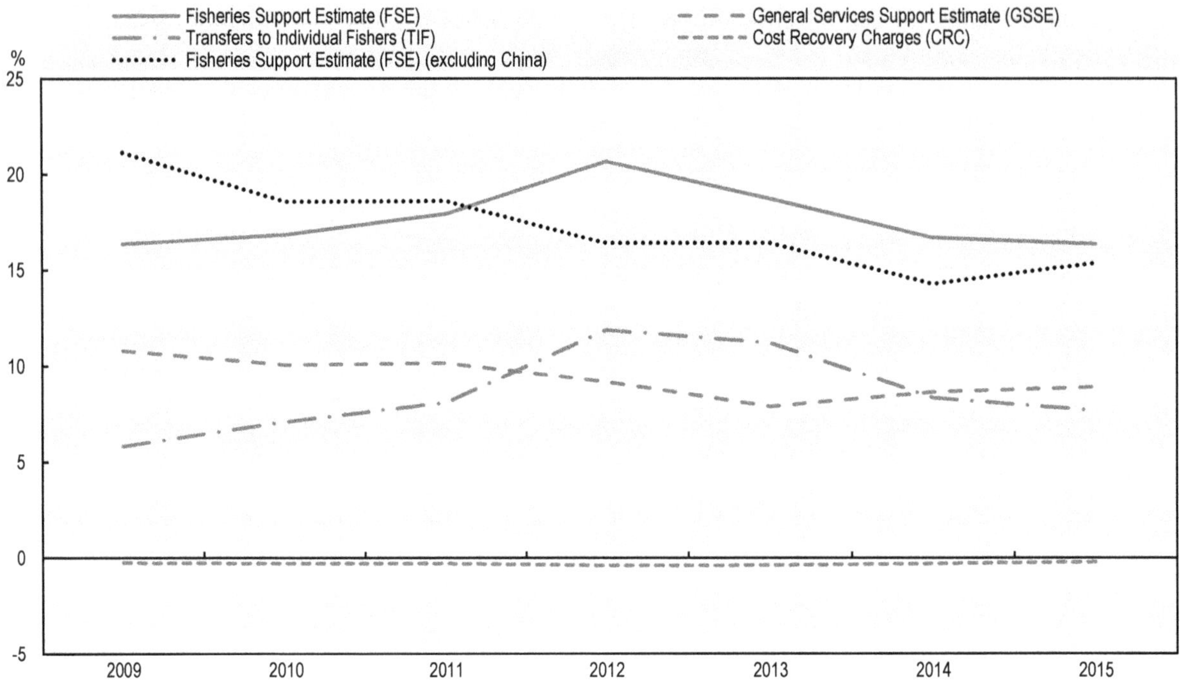

1. Missing figures for value of landings have been estimated using data from 'The 2016 Annual Economic Report on the EU Fishing Fleet' (Belgium, France, Lithuania, Spain) or in some cases simply carrying over the value of the previous year.
2. The GSSE numbers reported here include estimations for management and enforcement expenditures, where missing.
3. Missing 2015 FSE data for China have been estimated by carrying over 2014 values, except fuel subsidies which have been estimated based on announced goals
4. Further details on the estimation methods used are provided in the text.

Source: OECD Fisheries Database.

StatLink http://dx.doi.org/10.1787/888933627865

Box 3.1. OECD work on fisheries support helps inform trade negotiations

The Doha Ministerial Conference in 2005 launched negotiations to clarify and improve WTO disciplines on fisheries subsidies, including through a prohibition of certain forms of fisheries subsidies that contribute to overcapacity and overfishing. Those negotiations did not conclude. There has been new activity at the WTO on fisheries subsidies since 2016 and efforts appear to have intensified.

The OECD released a report on support to fisheries that was intended to improve the evidence-base for negotiations. Its preliminary analysis suggested that for a similar rate of support provided to the different categories of support in the FSE:

- Support directed at the use of variable inputs (such as for equipment, fuel or bait) is the most likely to increase fishing effort and the potential of overfishing, and the least likely to deliver real income benefits to fishers.
- Support directed at the use of fixed inputs (vessels and other durable investments) is the most likely to encourage expansion of capacity as this lowers the cost of investment. The benefits of this support disperses into factor values and benefits only factor owners at the time the support is initially put in place.
- Support based on fishers' income appears to provide the greatest benefit to fishers and is relatively less likely to increase capacity or fishing effort.
- Support to general services, a diverse category of policies including management, enforcement, infrastructure investments and R&D, appears to be the least likely to increase capacity or fishing effort.

The effect of support on capacity and fishing effort appears to be conditioned on, but not entirely mitigated by, the quality of the management system in place.

Source: OECD (2017), "Support to fisheries: Levels and impacts", *OECD Food, Agriculture and Fisheries Papers*, No. 103, OECD Publishing, Paris, www.dx.doi.org/10.1787/00287855-en.

Transfers to individual fishers (TIF) are provided in several forms

Transfers to individual fishers as a share of the budgetary FSE in OECD countries has been declining, and stood at 8% in 2015. This share jumps to 36% of budgetary FSE in 2009 and 47% in 2015 when Chinese data are included.

The TIFs reported in the FSE database are classified according to their implementation criteria, and cover forms of support that depend on the use of different factors of production, based on fishers' incomes, or payments to reduce productive capacity. Support that reduces variable input costs accounts for most of the reported TIF in the database.

The fuel subsidy in China is classified in the FSE as a budgetary transfer and not a tax benefit. Fishers receive a payment to reduce the cost of their fuel consumption that is based on the price of fuel. The subsidy covers 50% of the cost of diesel above RMB 3 870 (USD 615) per tonne and 100% of the cost above RMB 5 070 (USD 806) per tonne. The subsidy was introduced in 2006 and has expanded in value along with the quantity of fuel consumed, primarily by the distant water fleet. The Chinese government announced in its 13th Five-Year Plan to reduce the fuel subsidy to 40% of the 2014 value by 2019[6].

Australia had the highest rate of growth for transfers to fishers based on the reduction of productive capacity. Between the base period (2009-2011) and 2015 spending in this area increased by an average of 27% per year, predominantly due to the use of net-licence buyback schemes in 2015. In 2015, Australia was third behind **China** and **Spain** in terms of absolute spending on transfers based on the reduction of productive capacity. **China** was the largest provider of payments based on fixed capital formation (usually vessel construction or modernisation), followed by **Mexico**, **Indonesia** and **Spain**.

Transfers to individual fishers were below 5% of the value of landings in most cases in 2015 and less than 1% of the value of landings in 18 of the 34 countries covered. **New Zealand**, **Iceland** and **Lithuania** reported no budgetary transfers directed to individual fishers (Figure 3.2).[7] Only **Colombia** (44%), **Slovenia** (22%) and **China** (17%) provided support in the form of transfers to individual fishers that exceed 10% of their value of landings in 2015. In Slovenia's case, the relatively high amount results from a one-time payment made in 2015 under the European Fisheries Fund (EFF) "Socio-economic compensation for the management of the fishing fleet".

Figure 3.2. Budgetary support to fisheries as a share of value of landings, 2015

Disaggregated into Transfers to Individual Fishers (TIF) and General support to the sector (GSSE)

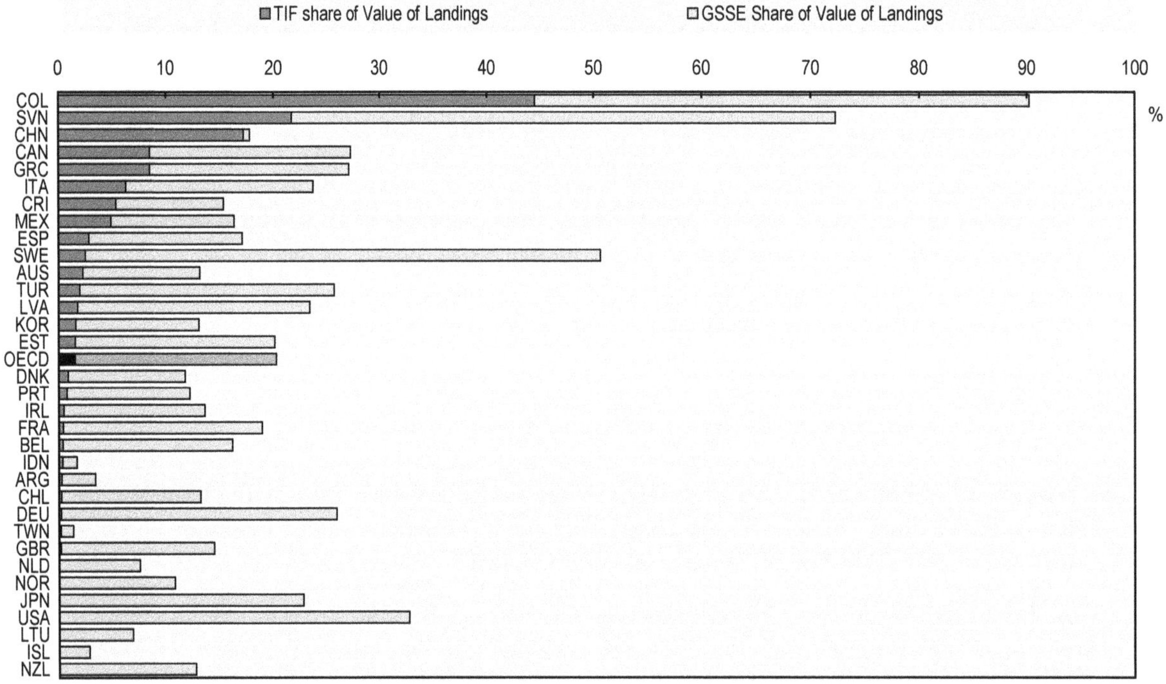

1. Missing figures for value of landings have been estimated using data from 'The 2016 Annual Economic Report on the EU Fishing Fleet' (Belgium, France, Lithuania, Spain) or in some cases simply carrying over the value of the previous year.
2. The GSSE numbers reported here include estimations for management and enforcement expenditures, where missing. Further details on the estimation method used are provided in the text.
3. Portugal and Mexico did not report any figure under the GSSE category so the numbers provided here reflect exclusively the estimations of management and enforcement expenditures.
4. Missing 2015 FSE data for China have been mostly estimated carrying over 2014 values, except fuel subsidies.

Source: OECD Fisheries Database.

StatLink http://dx.doi.org/10.1787/888933627884

There appears to be a minimum fixed cost associated with management and enforcement, as smaller fisheries tend to show these costs as a higher proportion of the value of landings. This may explain the situation in **Sweden** for example.

Decommissioning schemes still in widespread use, but the amount of support provided is in decline

Decommissioning schemes are policies to reduce or otherwise adjust fleet capacity through the scrapping of fishing vessels or retirement of fishing licenses. Such policies reached their height in the mid-2000s (OECD, $2006_{[10]}$). While the total amount of support provided to these programmes has been reduced over time, they remain in broad use illustrated by the number of countries reporting on them for this Review.

Fleet adjustment, through programmes designed to help reduce the overall number of vessels, fishing power, or both, is still an important tool. In the EU, the EMFF provides for such spending under Union Priority 1 on *Promoting environmentally sustainable, resource-efficient, innovative, competitive and knowledge-based fisheries* with 11 EU Member States intending to use decommissioning as part of their EMFF operational programmes for the 2014-2020 period. **China** has also made this a major emphasis of the fisheries development aspect of its 13th five-year plan. Similar policies have also been reported in **Chinese Taipei**, **Korea**, **Turkey** and **Australia.**

Support to general services is mostly used for enforcement measures

For most countries in the database, the relative importance of transfers to individual fishers has gradually decreased. This is partly a consequence of a decline in total spending in this area, but also because of an increase in absolute levels of support to general services between 2012 and 2015. The proportion of the total FSE in the form of general services policies has been increasing since 2009. After taking cost recovery charges into account, GSSE support made up 92% of total FSE transfers in 2015 (not including China), up from 83% in 2009 (Figure 3.3).[8]

Forms of support categorised as falling within general services include access to other exclusive economic zones (EEZs), infrastructure, marketing and promotion, community support, education and training, research and development, management of resources. When Chinese data is not included, expenditure on fisheries management programmes, surveillance and enforcement, and stock enhancement, accounts for the greatest part of general services support and accounted for 69% of all reported support (before cost recovery) in 2015. Provision of infrastructure (13%) and research and development (7%) accounted for most of the remaining reported support. China spends a relatively higher proportion of general services support on infrastructure, which changes the overall picture when this is taken into account.

About three-quarters of the support reported in the GSSE is for funding fisheries management or for monitoring, control and enforcement (MCS) (Figure 3.4). In the majority of countries MCS costs accounted for around 65% of total FSE spending in 2015. The exceptions to this are **Australia**, **Canada**, **Indonesia, Netherlands** and **Chinese Taipei**, who allocate a majority of their general support expenditures to fisheries management programmes. Other notable exceptions are **Sweden** and **Turkey** where 96% of general support is allocated to enforcement alone. In absolute terms, the **United States**, **Canada** and **Japan** report the most spending on stock enhancement, which can take the form of ecosystem restoration or restocking. In relative terms **Lithuania** allocates the greatest proportion of general support to restocking inland fisheries (35%).

Figure 3.3. General support to the sector, share and composition, 2009-11 vs 2015

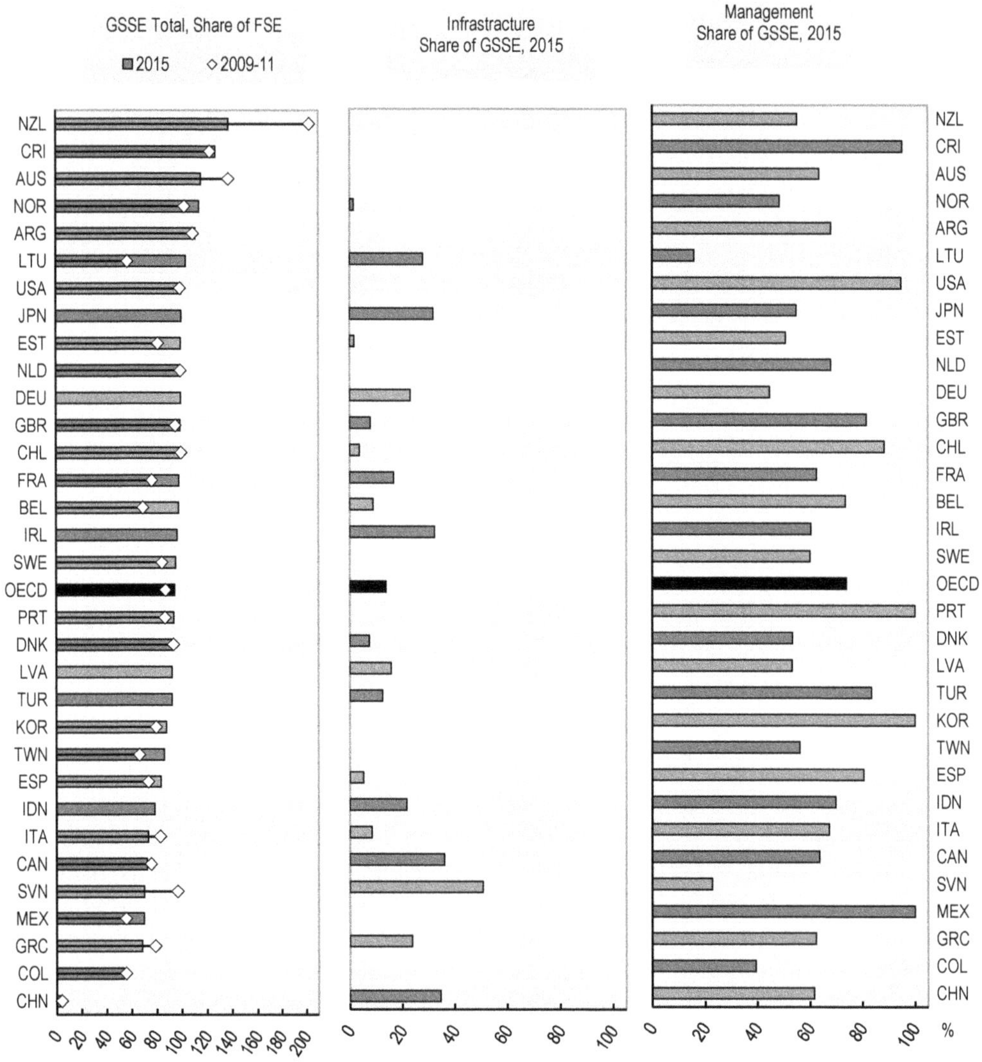

1. The GSSE numbers reported here include estimations for management and enforcement expenditures, where missing. Further details on the estimation method used are provided in the text.

2. Portugal and Mexico did not report any figure under the GSSE category so the numbers provided here reflect exclusively the estimations of management and enforcement expenditures.

3. Missing 2015 FSE data for China have been mostly estimated carrying over 2014 values, except fuel subsidies.

Source: OECD Fisheries Database.

StatLink http://dx.doi.org/10.1787/888933627903

Figure 3.4. Decomposition of management of resources, 2015

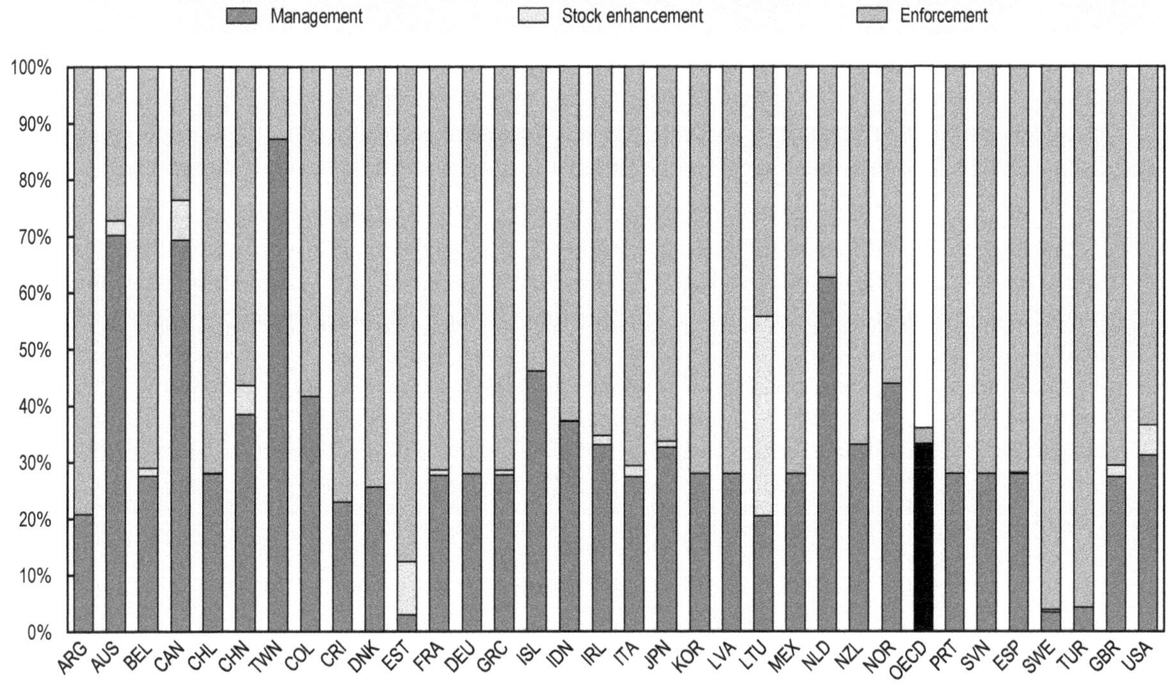

1. The numbers reported here include estimations for management and enforcement expenditures, where missing. Further details on the estimation method used are provided in the text.
2. Portugal and Mexico did not report any figure under the GSSE category so the numbers provided here reflect exclusively the estimations of management and enforcement expenditures.
3. Missing 2015 FSE data for China have mostly estimated carrying over 2014 values, except fuel subsidies.

Source: OECD Fisheries Database.

StatLink http://dx.doi.org/10.1787/888933627922

Cost recovery charges are small and not widely applied

Cost recovery charges (CRC) are typically levied in the context of offsetting the costs of providing GSSE support. As a proportion of the annual value of landings, these have fluctuated at an average of 2-3% in the countries that utilise CRC over the period 2009-15, in absolute terms the total value of cost recovery has increased by 3.8% per year over the same period. Of the 33 economies represented in the FSE database at some point between 2009 and 2015 only 10 reported the use of cost recovery charges. Some countries, such as **New Zealand** and **Iceland**, make substantial use of cost recovery and essentially collect the full cost of fisheries management in the form of charges to the sector.

Annual average rates of change in total FSE are mixed and uneven at the country level

The majority of countries have reduced reported support and in all but seven countries the absolute change is less than 10% per year (Figure 3.5). With the exception of China, the smallest changes have generally occurred in the largest fisheries. Support in China increased by 16% per year on average between the reference period and 2015 as spending on the fuel subsidy increased to its peak in 2014.

The largest annual average rate of change is seen in **Colombia** (+27% per year), whilst the largest decrease took place in **Iceland** (-22% per year). This is a result of the relatively low absolute levels of support they provide to their fishing industries, which make even small changes seem large in percentage terms. In 2015 **Iceland** had the lowest level of support as a share of landed value of all countries in the database; it also had the third lowest level of support in absolute terms. **Colombia** had the highest level of support as a share of landed value but it also had the fifth lowest absolute level of support in the same period and the trend in support is opposite that of the average.

The ratio of FSE to value of landings (Figure 3.2) puts the outcome of the apparently high growth rate observed for **New Zealand** into context. The explanation for this increase is that prior to 2012 New Zealand had not reported expenditures related to research. Research expenditures account for close to a third of New Zealand's FSE expenditures. This explains the significant yearly increase in New Zealand's FSE estimates from 2009 to 2015, so any increase in expenditure is likely to be significantly lower. Regardless of any growth in its FSE, New Zealand still has one of the lower ratios of support to value of landings of all OECD economies in the most recent period reported.

Figure 3.5. Evolution of budgetary support to fisheries, by country

Annual average change, 2009-11 to 2015

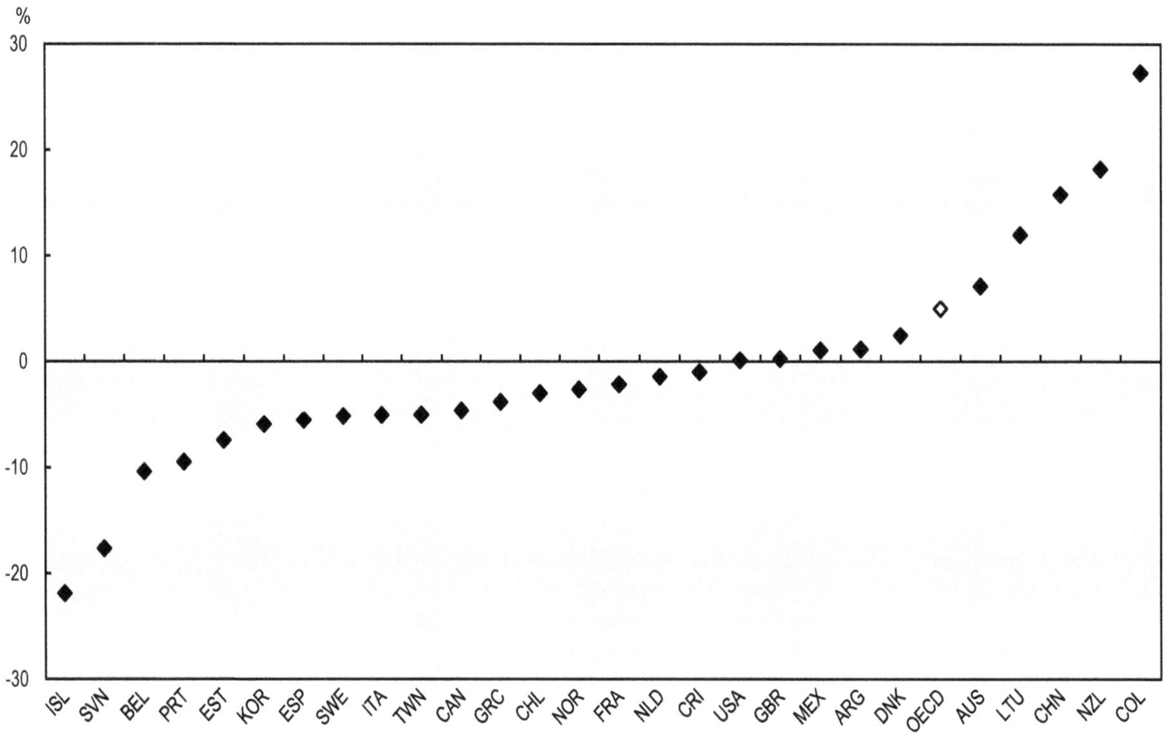

1. The numbers reported here include estimations for management and enforcement expenditures, where missing. Further details on the estimation methods used are provided in the text.
2. Missing 2015 FSE data for China have been mostly estimated carrying over 2014 values, except fuel subsidies.
Source: OECD Fisheries Database.

StatLink http://dx.doi.org/10.1787/888933627941

Most countries have reduced the amount of support provided directly to individual fishers

Eighteen countries have reduced the amount of support provided to individual fishers between the 2009-11 reference period and 2015. (Figure 3.6—those in the left half). Of these, nine have also reduced the amount of support to general services in the same period. Reductions in TIF have generally been proportionally larger than those in the GSSE, but this is largely driven by the relative difference in the absolute sizes of the two forms of support. Nine countries increased spending on TIF, of which four increased both TIF and GSSE (**Colombia**, **China, Australia** and **Denmark**).

Figure 3.6. Change in budgetary support to fisheries by component, by country

Annual average change, 2009-11 to 2015

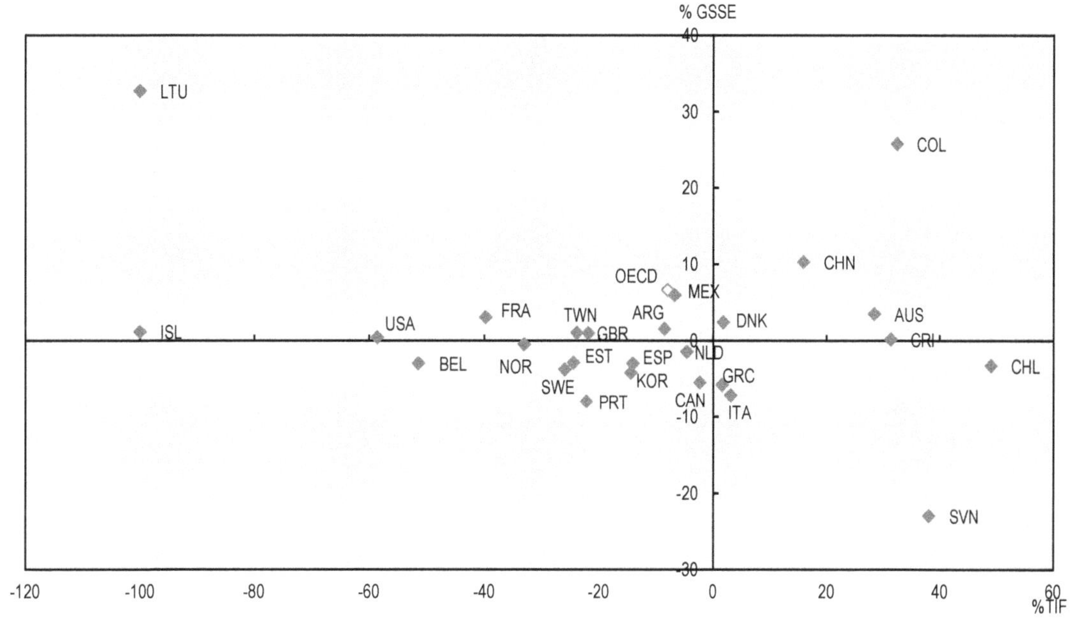

1. The GSSE numbers reported here include estimations for management and enforcement expenditures, where missing. Further details on the estimation methods used are provided in the text.
2. Portugal and Mexico did not report any figure under the GSSE category so the numbers provided here for GSSE changes reflect exclusively the estimations of management and enforcement expenditures.
3. Missing 2015 FSE data for China have been mostly estimated carrying over 2014 values, except fuel subsidies.
Source: OECD Fisheries Database.

StatLink http://dx.doi.org/10.1787/888933627960

Notes

[1] The fuel subsidy in China is a budgetary transfer, not a tax concession.

[2] Not all economies participating in this Review participate in the FSE database. Currently the following countries or economies are included; Australia, Belgium, Canada, Chile, Denmark, Estonia, France, Germany, Greece, Iceland, Ireland, Italy, Japan, Korea, Latvia, Mexico, Netherlands, New Zealand, Norway, Portugal, Slovenia, Spain, Sweden, Turkey, United Kingdom, United States, Colombia, Costa Rica, Lithuania, China (People's Republic of), Indonesia, Argentina, Chinese Taipei.

[3] Support considered as falling within the category of general services includes access to other EEZs, infrastructure, marketing and promotion, community support, education and training, research and development, management of resources.

[4] Transfers that support fishers individually include support that reduces variable or fixed costs, maintains income, and payments to reduce capacity

[5] In a small number of cases data on the value of landings was not available for 2015. Estimated management and enforcement expenditures in 2015 were therefore calculated using the value of landings for that country in 2014. Indicators where the value of landings is used to contextualise the level of support also use value of landings from 2014 if it was not available for 2015. This is under the assumption that value of landings will have not changed significantly from one year to the next. In all cases, if the data used in an indicator or associated figure is not for 2015 this is stated in the text or in a note appended to the figure. Following this approach the indicators presented here provide the most up-to-date and complete picture currently possible of support to fisheries over the period 2009-15.

[6] This reduction amount is used to (linearly) estimate the change in the subsidy between 2014, the last year available, and 2015.

[7] EU Member State support is heavily influenced in scope and volume by EU structural support (EFF, EMFF).

[8] Improved policy coverage in both the TIF and GSSE components of the FSE could change this picture.

References

Environment and Climate Change Canada (2017), *Canadian Environmental Sustainability Indicators: Status of Major Fish Stocks*, http://www.ec.gc.ca/indicateurs-indicators/default.asp?lang=en&n=1BCD421B-1. [3]

Environment and Climate Change Canada (2017), *Canadian Environmental Sustainability Indicators: Sustainable Fish Harvest*, http://www.ec.gc.ca/indicateurs-indicators/default.asp?lang=en&n=893AB9F4-1. [4]

Fisheries And Oceans Canada (2016), *Sustainability Survey for Fisheries*, http://, http://www.dfo-mpo.gc.ca/reports-rapports/regs/sff-cpd/survey-sondage/index-en.html. [5]

FRDC (2017), *Status of Australian Fish Stocks Reports*, Fisheries Research and Development Corporation, http://fish.gov.au/ReportStock?kw=&page=1&sort=LatestFirst. [7]

Hafrannsóknastofnun (2016), *State of Marine Stocks in Icelandic Waters 2015/2016 and Prospects for the Quota Year 2016/2017*, Hafrannsóknastofnun (Marine Research Institute), [6]

http://www.hafro.is/Astand/2016/fjolrit_185.pdf.

MAFF (2016), *FY 2015 Trends in Fisheries*, Ministry of Agriculture, Forestry and Fisheries, http://www.maff.go.jp/e/data/publish/attach/pdf/index-49.pdf. [2]

Martini, R. (2012), "Fuel Tax Concessions in the Fisheries Sector", *OECD Food, Agriculture and Fisheries Papers*, No. 56, OECD Publishing, Paris, http://dx.doi.org/10.1787/5k9bdccqft30-en. [9]

NOAA (2017), *Status of Fish Stocks 2016*, NOAA Fisheries, http://www.nmfs.noaa.gov/sfa/fisheries_eco/status_of_fisheries/archive/2016/status-of-stocks-2016-web.pdf. [1]

OECD (2006), *Financial Support to Fisheries: Implications for Sustainable Development*, OECD Publishing, Paris, http://dx.doi.org/10.1787/9789264036642-en. [10]

OECD (2017), *Marine Protected Areas: Economics, Management and Effective Policy Mixes*, OECD Publishing, Paris, http://dx.doi.org/10.1787/9789264276208-en. [8]

Chapter 4. OECD country and non-OECD economy snapshots

This chapter provides additional country-level data and details on institutions and policies based on contributions by participating countries and economies.

The statistical data for Israel are supplied by and under the responsibility of the relevant Israeli authorities. The use of such data by the OECD is without prejudice to the status of the Golan Heights, East Jerusalem and Israeli settlements in the West Bank under the terms of international law.

Argentina

Developments in fisheries and aquaculture policy, 2015-16

- In 2016, Argentina started developing a strategy to implement ecosystem based fisheries management with support from the Global Environment Facility programme.
- In 2015, the National Plan of Action to reduce the number of interactions with marine mammals was approved. This came after earlier approved action plans to protect seabirds and sharks.
- Since 2015, a new system of on-board cameras has been implemented to assure better compliance with regulations to strengthen control and surveillance over the fisheries sector.

Fisheries and aquaculture statistics in Argentina

Value of production

in global context

USD 1 897 million	1.5%	0.02%
Total value of fisheries and aquaculture	*Share of global fisheries landed value*	*Share of global aquaculture production value*

Trends in production value and trade

Total production value in USD increased at an annual average rate of 2.9% between 2011 and 2015

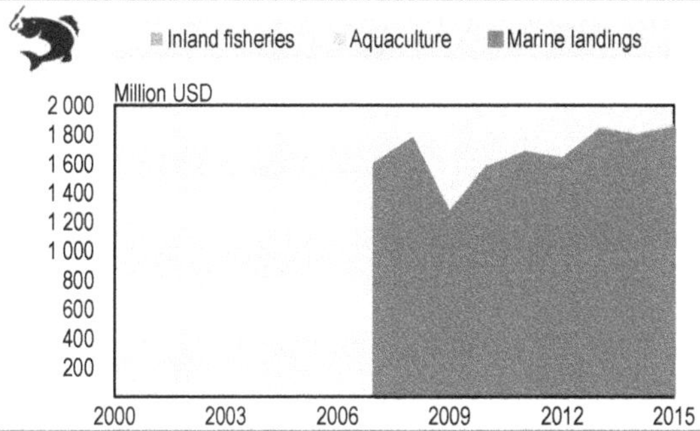

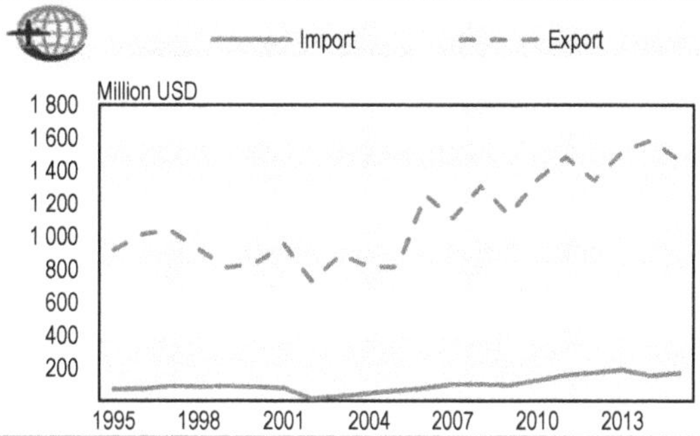

Employment and fleet

Employment in capture fisheries decreased by 0.3% between 2014 and 2015 to 20 866 persons

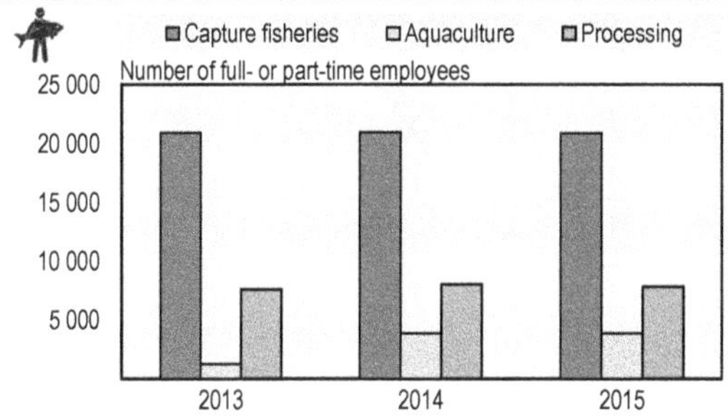

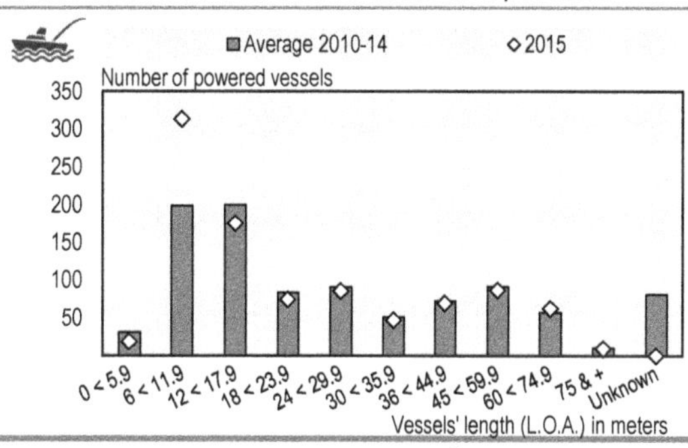

Government support and resource management

76% of total budgetary support was directed to management of resources in 2015

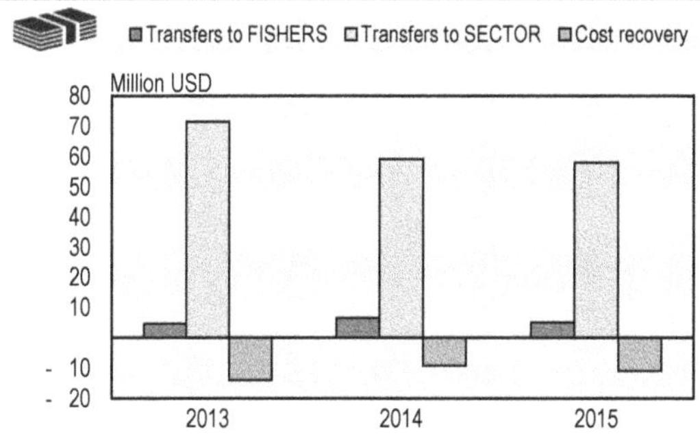

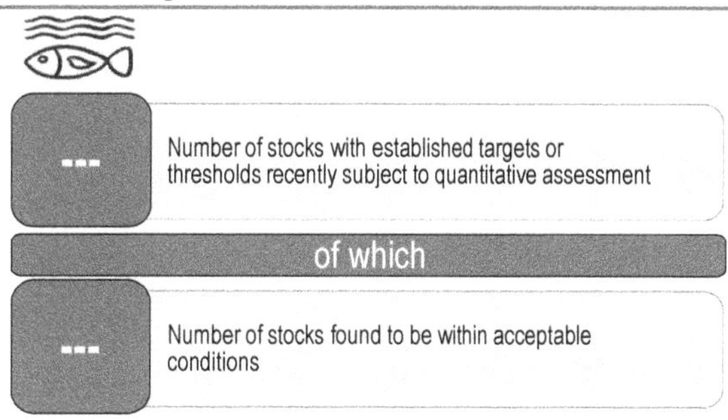

Note: Reference year is 2015. The 'Transfers to SECTOR' numbers reported here include estimations for management and enforcement expenditures, where missing.

Source: OECD Fisheries Database. World's total values of production for catch and aquaculture were sourced from the *FAO Yearbook, Fishery and Aquaculture Statistics, 2015*. Both numbers have been corrected adding the value of seaweed production; aquaculture has also been adjusted to account for the value of Chinese production reported to the OECD.

Australia

Developments in fisheries and aquaculture policy, 2015-16

- Under its newly developed National Aquaculture Strategy, Australia aims to double the value of its existing AUD 1.2 billion (USD 0.9 billion) aquaculture industry by 2027.
- In March 2017, a national seafood industry representative body was established with AUD 535 000 of seed funding from the Australian Government. The new body will be the voice of the industry at the national level and promote a viable and sustainable seafood industry with representation from the wild-catch, aquaculture and post-harvest sectors.
- In November 2015, the Australian Government published a framework for conducting recreational fishing surveys and committed to undertaking regular surveys to collect data on the social and economic impact of recreational fishing.
- Australia signed three international agreements in 2015 and 2016 that will reduce or eliminate tariffs on the trade of fisheries products.
- Stocks managed solely by the Australian Government were found to be free of overfishing in the annual Fishery Status Report 2016.

Fisheries and aquaculture statistics in Australia

Value of production

in global context

USD 2 099 million	1.0%	0.4%
Total value of fisheries and aquaculture	*Share of global fisheries landed value*	*Share of global aquaculture production value*

Trends in production value and trade

Total production value in USD decreased at an annual average rate of 2.7% between 2011 and 2015

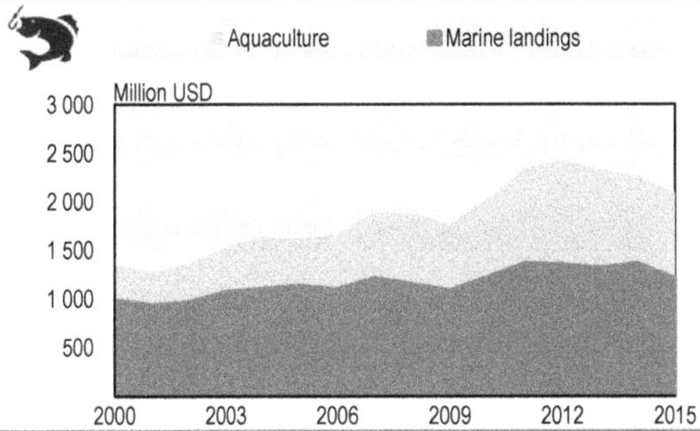

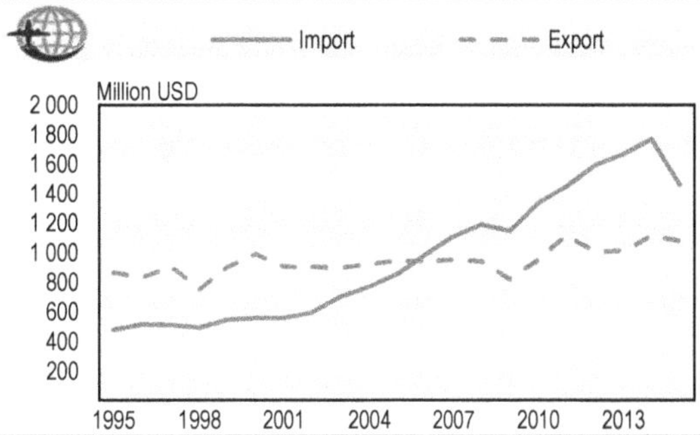

Employment and fleet

Employment in capture fisheries increased by 101.6% between 2014 and 2015 to 7 225 persons

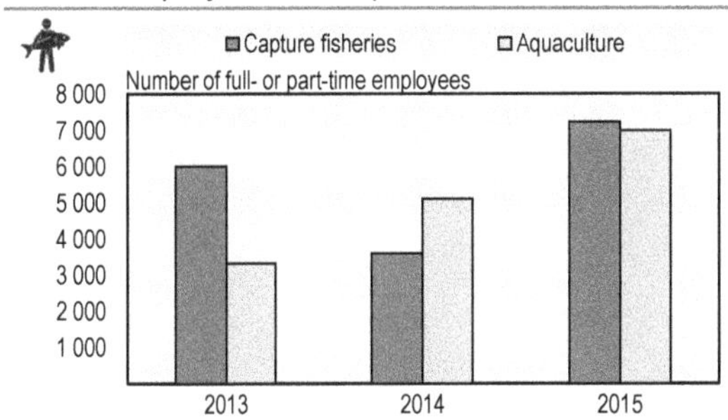

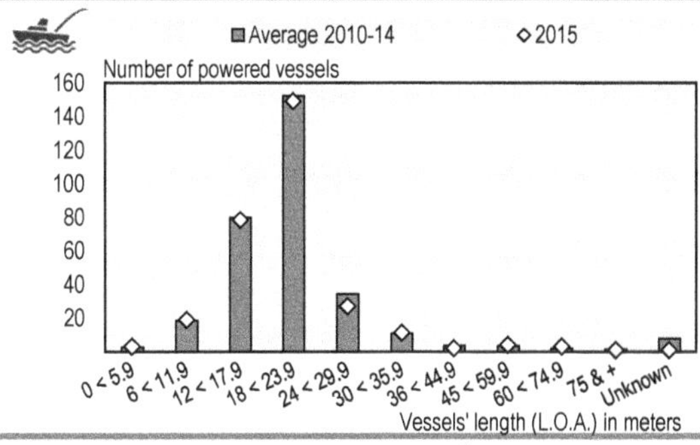

Government support and resource management

73% of total budgetary support was directed to management of resources in 2015

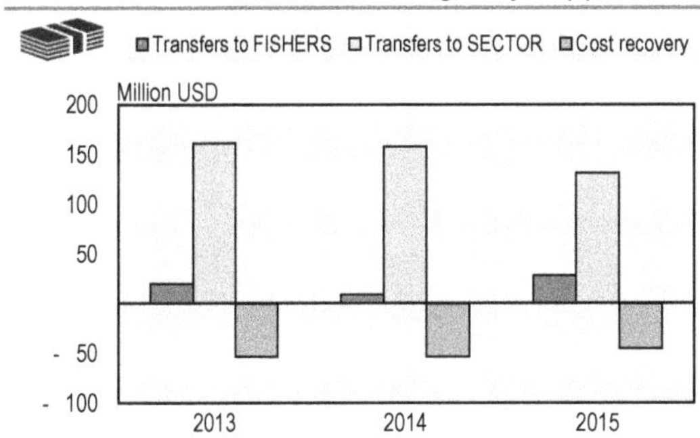

232 Number of stocks with established targets or thresholds recently subject to quantitative assessment

of which

201 Number found to have acceptable status with respect to targets or thresholds

Note: Reference year is 2015, except for stock assessment data which may cover more than one year.
Source: OECD Fisheries Database. World's total values of production for catch and aquaculture were sourced from the *FAO Yearbook, Fishery and Aquaculture Statistics, 2015*. Both numbers have been corrected adding the value of seaweed production; aquaculture has also been adjusted to account for the value of Chinese production reported to the OECD. Fish stocks data were directly reported by the country and are available in the Status of Australian Fish Stocks Reports.

Belgium

Developments in fisheries and aquaculture policy, 2015-16

- Belgium ended its overcapacity period by decreasing the number of vessels. The Belgian fleet counted 72 vessels in 2016, four less than in the previous year and a decrease from 100 vessels in 2008.
- Belgium's Operational Programme (OP) contributes to increasing the sustainability of the fisheries and aquaculture sectors, from both an economic and environmental standpoint. Investment and increases in selectivity drive the core strategy for fisheries. For aquaculture, increased investments in production capacity combined with greater exchange of knowledge and research aim to increase production.
- The Belgian government actively supports work on a strategy to operationalise a maritime spatial planning exercise in the European Union.

Fisheries and aquaculture statistics in Belgium

Value of production

in global context

USD 91 million	0.07%	0.000%
Total value of fisheries and aquaculture	*Share of global fisheries landed value*	*Share of global aquaculture production value*

Trends in production value and trade

Total production value in USD decreased at an annual average rate of 4.8% between 2011 and 2015

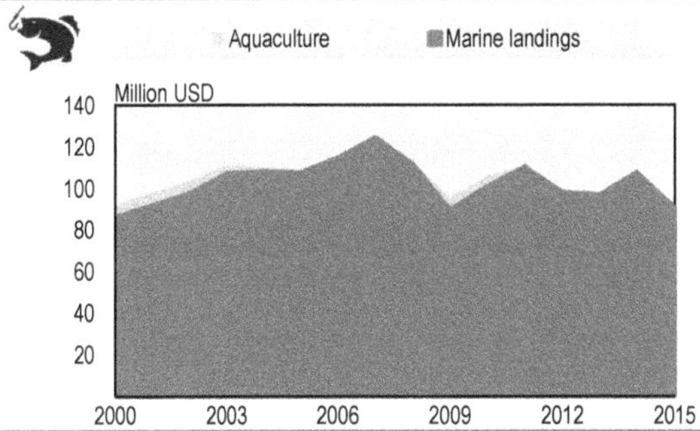

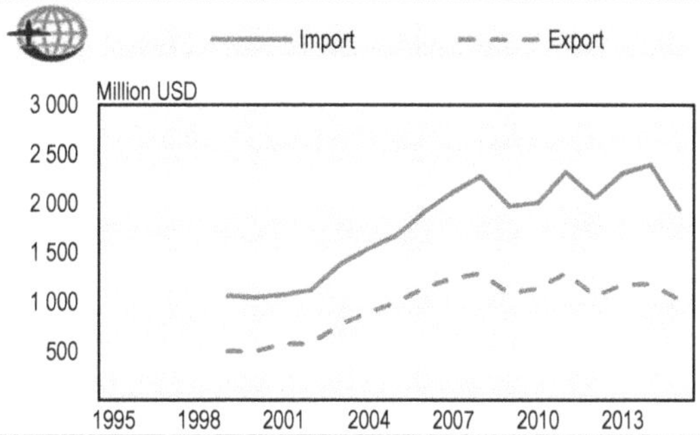

Employment and fleet

Employment in capture fisheries decreased by 12.5% between 2014 and 2015 to 499 persons

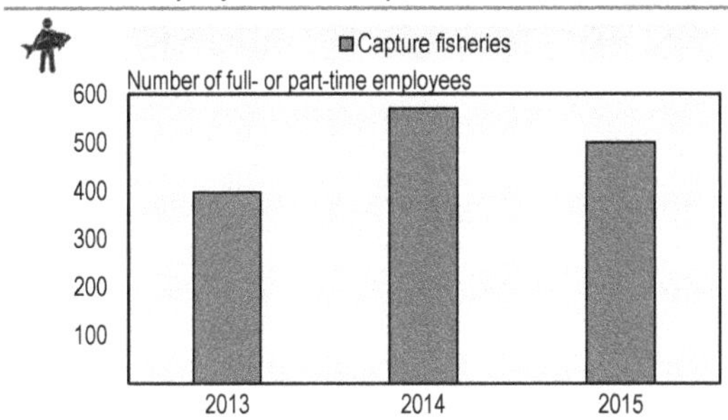

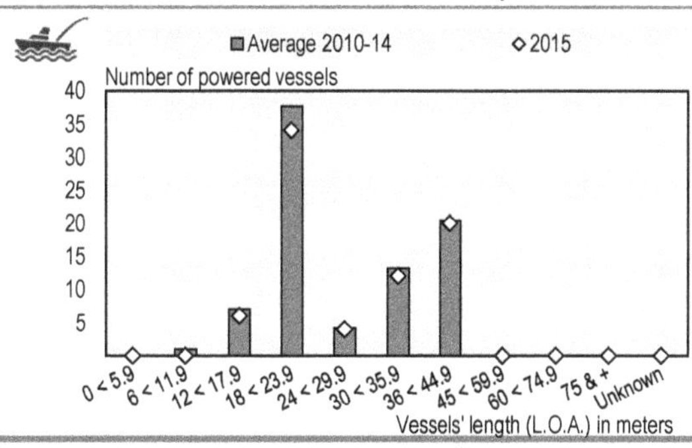

Government support and resource management

72% of total budgetary support was directed to management of resources in 2015

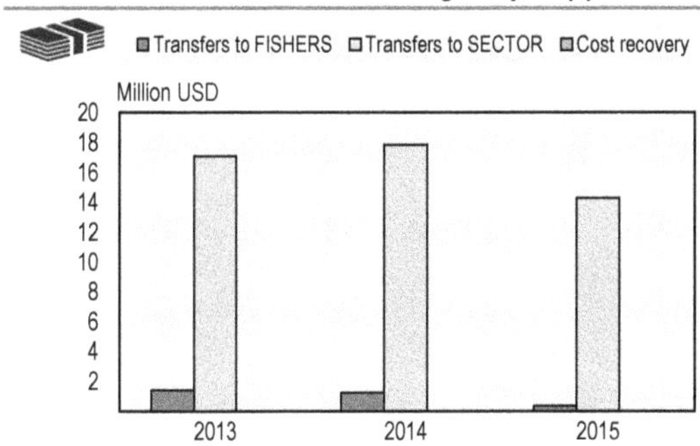

For information on EU fish stocks please refer to the EU country chapter

Note: Reference year is 2015. The 'Transfers to SECTOR' numbers reported here include estimations for management and enforcement expenditures, where missing.

Source: OECD Fisheries Database. World's total values of production for catch and aquaculture were sourced from the *FAO Yearbook, Fishery and Aquaculture Statistics, 2015*. Both numbers have been corrected adding the value of seaweed production; aquaculture has also been adjusted to account for the value of Chinese production reported to the OECD.

Canada

Developments in fisheries and aquaculture policy, 2015-16

- Results of the Sustainability Survey for Fisheries (SSF) were published for the first time in 2016, reporting on progress towards implementation of the policies within the Sustainable Fisheries Framework (SFF). While most fisheries have been brought into alignment with multiple elements of the SSF, additional work remains.
- On World Oceans Day, June 8, 2016, Canada announced its commitment to increase marine conservation from approximately 1% of Canada's vast marine and coastal areas to 5% by 2017 and 10% by 2020. In October of 2017, Canada announced that it has achieved the 5% target.
- In 2016, a review of the 2012 Fisheries Act amendments began – one of several environmental legislative and regulatory reviews concurrently underway. The review is focused on gaps in protection under the current legislation and on how to incorporate modern safeguards to protect and conserve fish and fish habitats.
- In July 2015, after four years of development, the Aquaculture Activities Regulations (AAR) – the first national aquaculture regulations – came into effect. Under authority of the Fisheries Act, the AAR provides greater environmental assurances and regulatory stability for businesses.

Fisheries and aquaculture statistics in Canada

Value of production

in global context

USD 3 307 million	2.1%	0.4%
Total value of fisheries and aquaculture	*Share of global fisheries landed value*	*Share of global aquaculture production value*

Trends in production value and trade

Total production value in USD increased at an annual average rate of 2.2% between 2011 and 2015

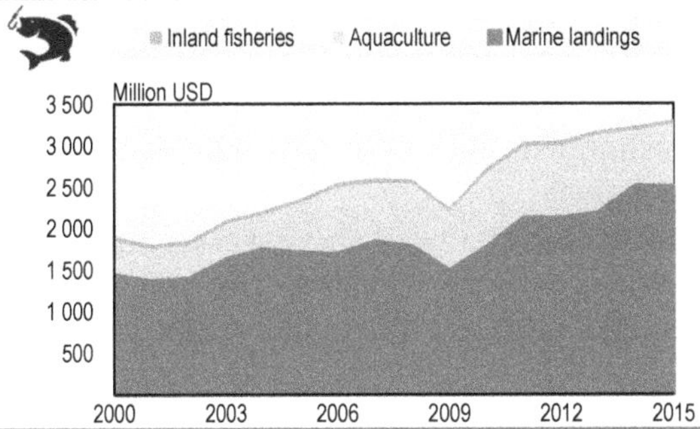

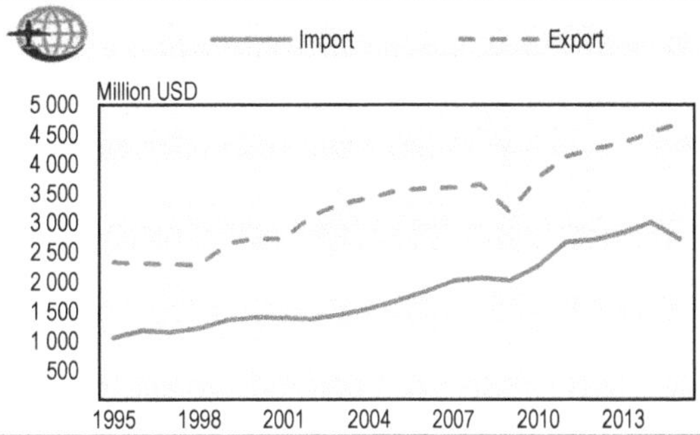

Employment and fleet

Employment in capture fisheries decreased by 2.4% between 2014 and 2015 to 42 507 persons

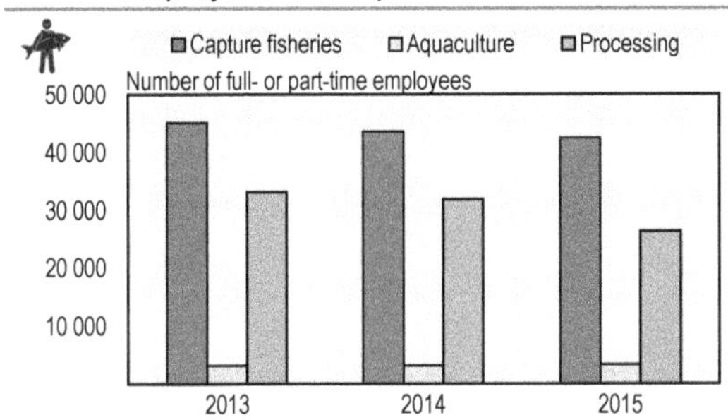

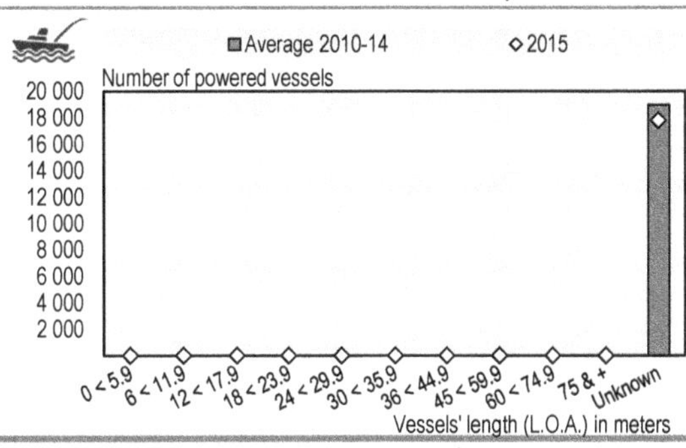

Government support and resource management

46% of total budgetary support was directed to management of resources in 2015

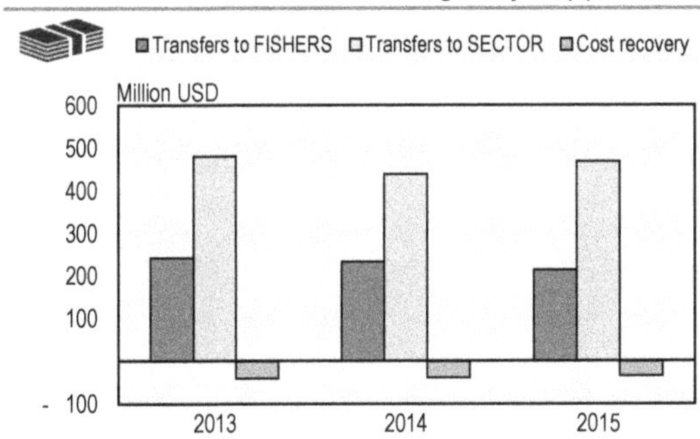

159 Number of stocks with established targets or thresholds recently subject to quantitative assessment

of which

76 Number where stock status was both in the healthy zone and harvest rate was either at or below the removal reference or deemed within approved levels

Note: Reference year is 2015, except for stock assessment data which may cover more than one year.
Source: OECD Fisheries Database. World's total values of production for catch and aquaculture were sourced from the *FAO Yearbook, Fishery and Aquaculture Statistics, 2015*. Both numbers have been corrected adding the value of seaweed production; aquaculture has also been adjusted to account for the value of Chinese production reported to the OECD. Fish stocks data were sourced from Canada's 2015 Sustainability Survey for Fisheries. A stock is considered to be in the healthy zone if its biomass, or the index of the biomass, is higher than its Upper Stock Reference Point (e.g. higher than 80% of B_{MSY}).

Chile

Developments in fisheries and aquaculture policy, 2015-16

- Chile's legislative framework in for 2014-18 includes a portfolio of new regulations that focus on the development of artisanal fishing and small-scale aquaculture. The establishment of the National Institute of Sustainable Development of Artisanal Fisheries and Small-Scale Aquaculture is close to final approval by the National Congress.
- Action has been taken in recent years to tackle illegal fishing in Chile's waters. The National Congress is discussing a bill which would strengthen the authority of the National Fisheries and Aquaculture Service, an executive body that supervises compliance with fisheries and aquaculture regulations.
- Chile is investing in initiatives that reduce the vulnerability of the fisheries and aquaculture sectors to climate change. Several measures are underway that aim at better adaptive capacity, including the Climate Change Adaptation Plan for Fisheries and Aquaculture approved in 2015.
- In 2016, Chile's fisheries and aquaculture suffered form an unusually intense harmful algae bloom. Wide areas were closed for production by sanitary authorities. An extremely high concentration of toxins in harvested fish and cultivated mussels months after the outbreak resulted in a series of modifications in sanitary control and surveillance processes.

Fisheries and aquaculture statistics in Chile

Value of production

in global context

USD 7 973 million	0.9%	3.2%
Total value of fisheries and aquaculture	*Share of global fisheries landed value*	*Share of global aquaculture production value*

Trends in production value and trade

Total production value in USD increased at an annual average rate of 0.6% between 2011 and 2015

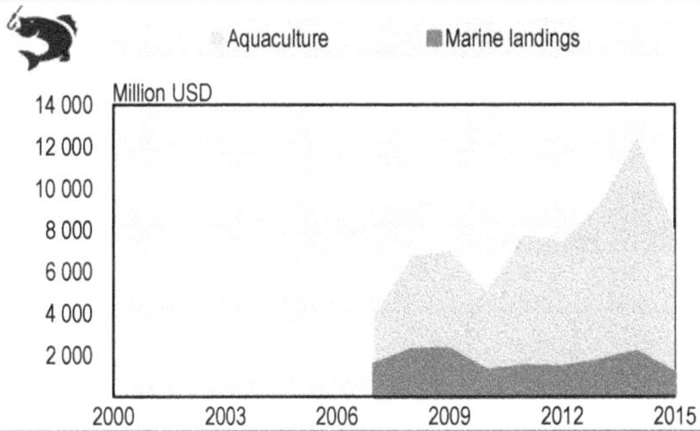

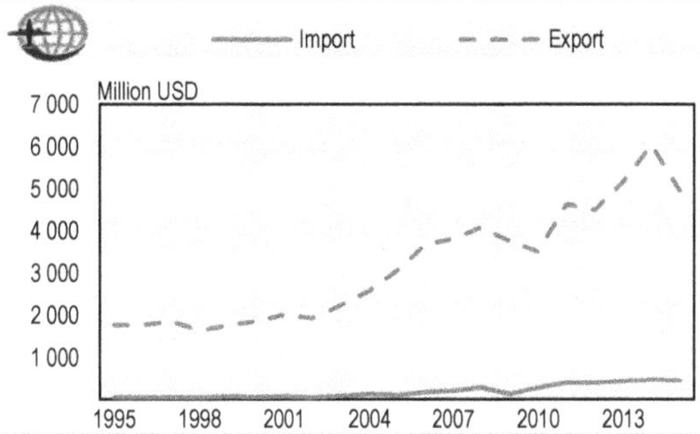

Employment and fleet

Employment in capture fisheries increased by 3.5% between 2014 and 2015 to 96 806 persons

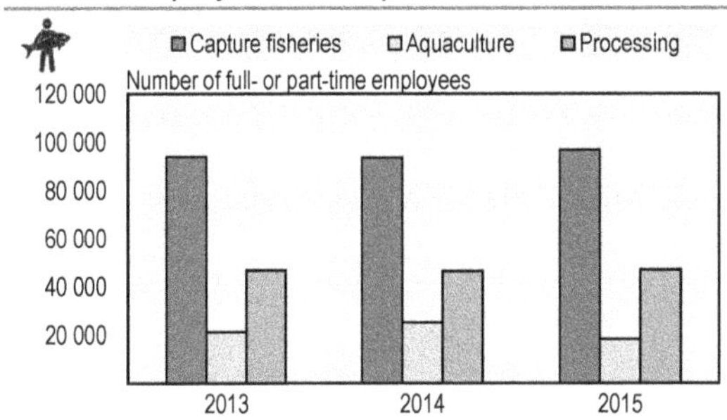

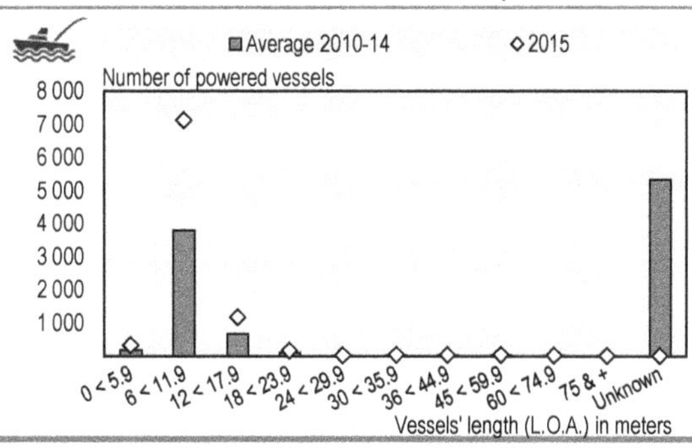

Government support and resource management

87% of total budgetary support was directed to management of resources in 2015

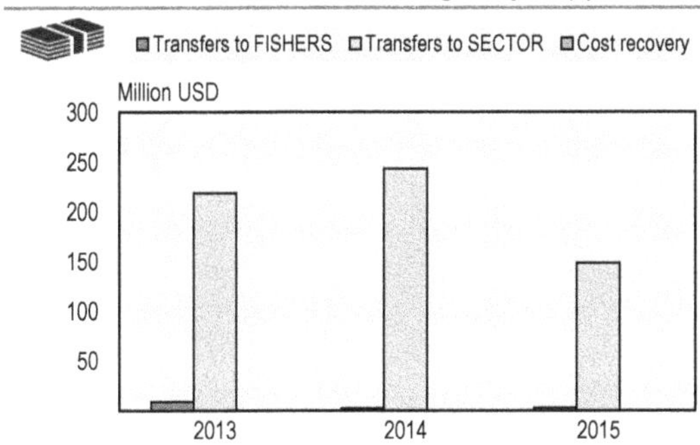

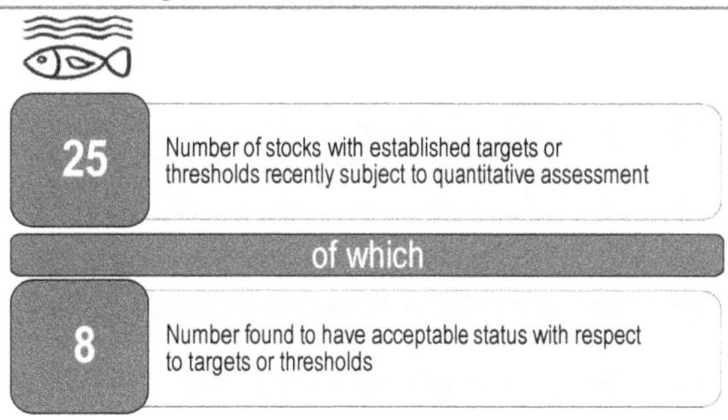

Note: Reference year is 2015, except for stock assessment data which may cover more than one year. The 'Transfers to SECTOR' numbers reported here include estimations for management and enforcement expenditures, where missing.

Source: OECD Fisheries Database. World's total values of production for catch and aquaculture were sourced from the *FAO Yearbook, Fishery and Aquaculture Statistics, 2015*. Both numbers have been corrected adding the value of seaweed production; aquaculture has also been adjusted to account for the value of Chinese production reported to the OECD. Fish stocks data were directly reported by the country.

People's Republic of China

Developments in fisheries and aquaculture policy, 2015-16

- The 13th five year plan (2016-2020) envisions a more sustainable and market-oriented fishery in the People's Republic of China (hereafter "China"), continuing a shift away from past emphasis on increasing production.
- The diesel fuel subsidy will be progressively reduced to 40% of the 2014 amount by 2019, with the funds shifted to fisheries resources conservation, vessel modernisation and infrastructure.
- Artificial restocking, artificial reefs and seasonal closures are the main measures used for restoration of fish stocks.
- Distant water fisheries are the main source of growth in capture fisheries landings, while aquaculture is the main source of growth in total fish production.

Fisheries and aquaculture statistics in China

Value of production

in global context

USD 170 474 million	32.0%	61.2%
Total value of fisheries and aquaculture	*Share of global fisheries landed value*	*Share of global aquaculture production value*

Trends in production value and trade

Total production value in USD increased at an annual average rate of 10% between 2011 and 2015

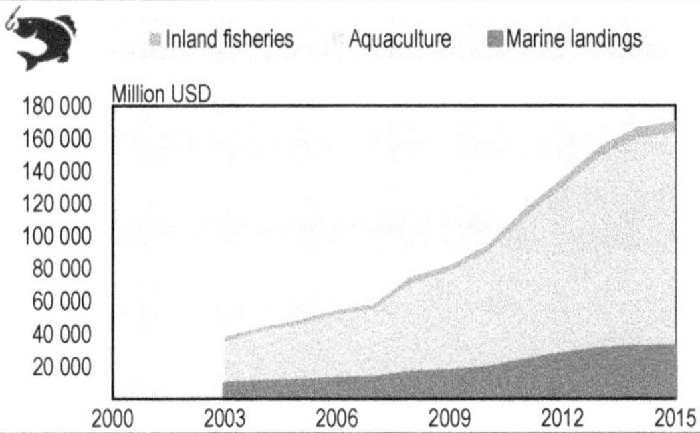

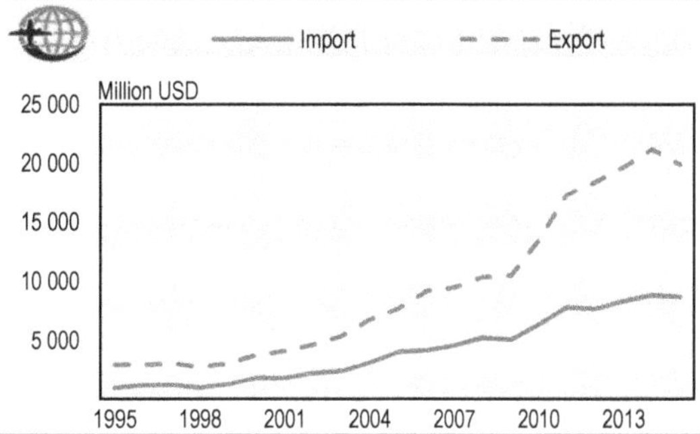

Employment and fleet

Employment in capture fisheries decreased by 3.5% between 2014 and 2015 to 1 753 618 persons

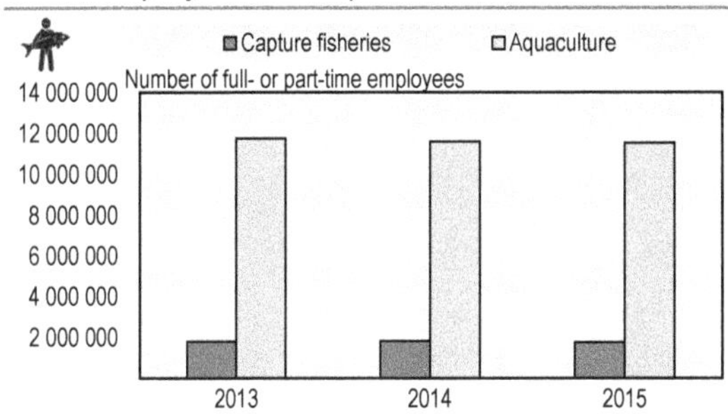

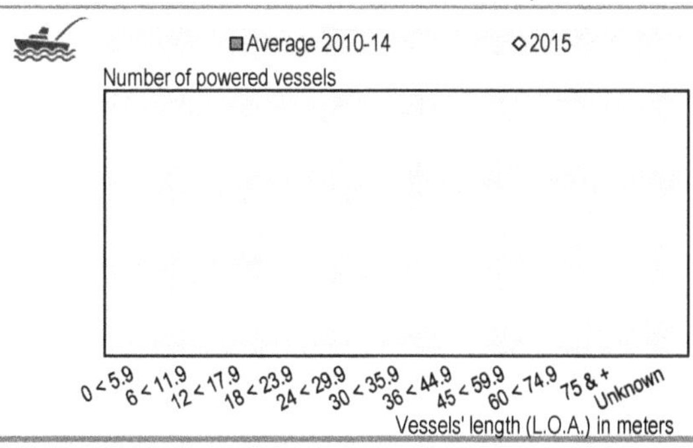

Government support and resource management

2% of total budgetary support was directed to management of resources in 2015

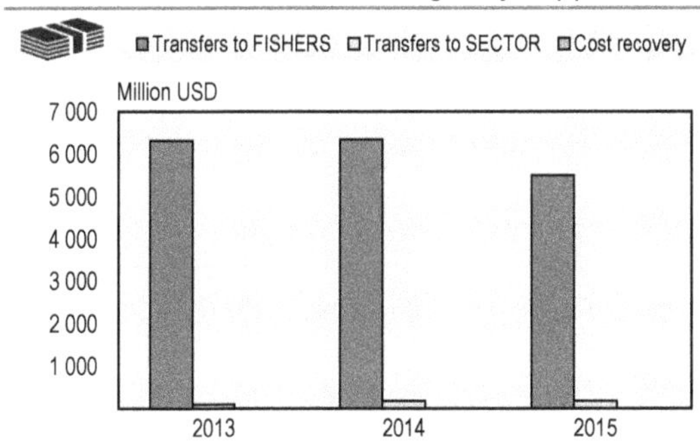

---	Number of stocks with established targets or thresholds recently subject to quantitative assessment
of which	
---	Number of stocks found to be within acceptable conditions

Note: Reference year is 2015.

Source: OECD Fisheries Database. World's total values of production for catch and aquaculture were sourced from the *FAO Yearbook, Fishery and Aquaculture Statistics, 2015*. Both numbers have been corrected adding the value of seaweed production; aquaculture has also been adjusted to account for the value of Chinese production reported to the OECD. Data on aquaculture production were sourced from the China Statistical Yearbook 2016; differences with FAO total figures are under review. 2015 budgetary support data have been estimated using 2014 values.

Colombia

Developments in fisheries and aquaculture policy, 2015-16

- In 2015 and 2016, progress has been made with the implementation of the "Comprehensive policy for the development of sustainable fisheries in Colombia". At the same time, the legislative process has advanced on the introduction of the law on illegal fishing activities.
- In 2016, Colombia allocated approximately USD 14 million to the development of fisheries and aquaculture industry, nearly triple (considering values in USD) compared with 2011. Most of the budget is earmarked for research and development.
- Colombia continues to struggle with environmental issues affecting the availability of fisheries resources. The causes include poor water quality due to pollution from inland activities, increased sedimentation following deforestation, climate variability and overfishing.

Fisheries and aquaculture statistics in Colombia

Value of production

in global context

USD 259 million	---	0.1%
Total value of fisheries and aquaculture	*Share of global fisheries landed value*	*Share of global aquaculture production value*

Trends in production value and trade

Total production value in USD decreased at an annual average rate of 0.3% between 2011 and 2015

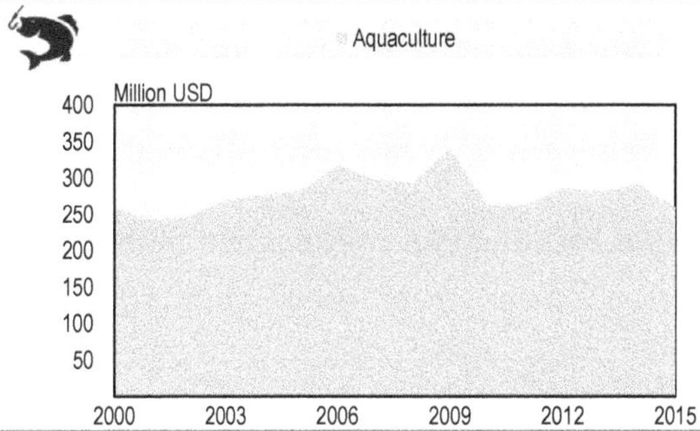

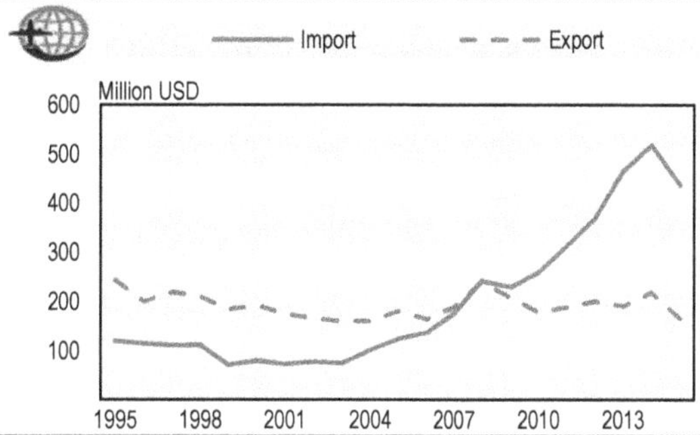

Employment and fleet

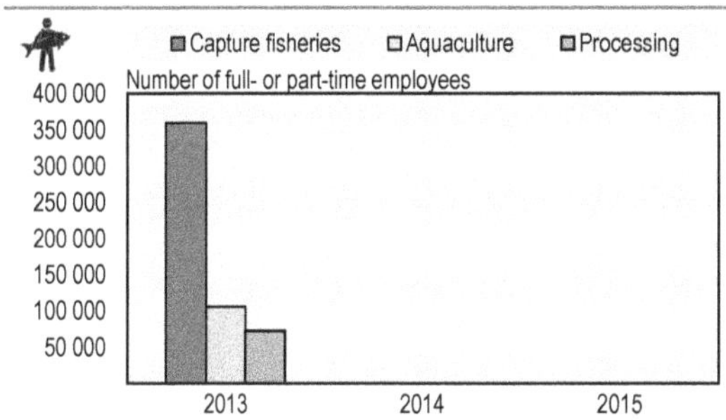

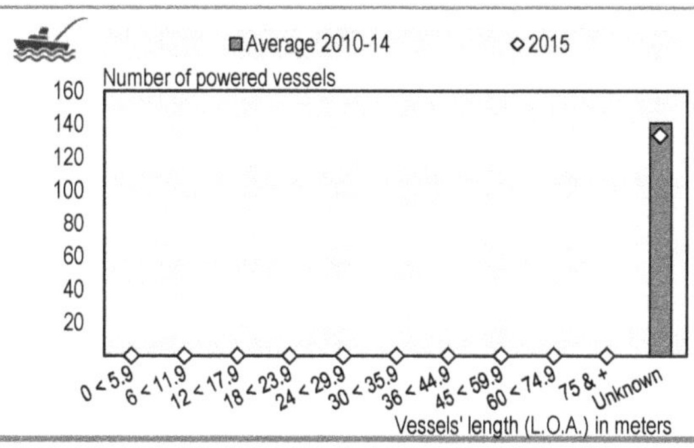

Government support and resource management

21% of total budgetary support was directed to management of resources in 2015

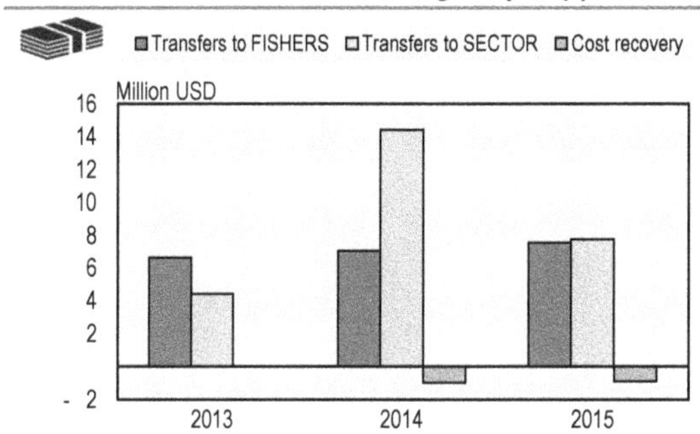

21 Number of stocks with established targets or thresholds recently subject to quantitative assessment

of which

8 Number found to have acceptable status with respect to targets or thresholds

Note: Reference year is 2015, except for stock assessment data which may cover more than one year.
Source: OECD Fisheries Database. World's total values of production for catch and aquaculture were sourced from the *FAO Yearbook, Fishery and Aquaculture Statistics, 2015*. Both numbers have been corrected adding the value of seaweed production; aquaculture has also been adjusted to account for the value of Chinese production reported to the OECD. Fish stocks data were directly reported by the country.

Czech Republic

Developments in fisheries and aquaculture policy, 2015-16

- The Czech Republic will receive support from the European Maritime and Fisheries Fund (EMFF) according to European Union (EU) regulations. The funds will support aquaculture in the aim of making the sector more productive without compromising sustainability and conservation goals. The total budget for the Operational Programme (OP) is approximately EUR 41 million (USD 55 million) for the period 2014-20.
- Additional government financial transfers, of approximately EUR 5 million (USD 5.5 million) annually in the last years, represent a broad spectrum of the supports for freshwater aquaculture.
- The government is also funding actions aimed at boosting fish consumption in the country; these measures include investments in improved market organisation and marketing to consumers.

Fisheries and aquaculture statistics in the Czech Republic

Value of production

in global context

USD 47 million	---	0.02%
Total value of fisheries and aquaculture	*Share of global fisheries landed value*	*Share of global aquaculture production value*

Trends in production value and trade

Total production value in USD decreased at an annual average rate of 8.9% between 2011 and 2015

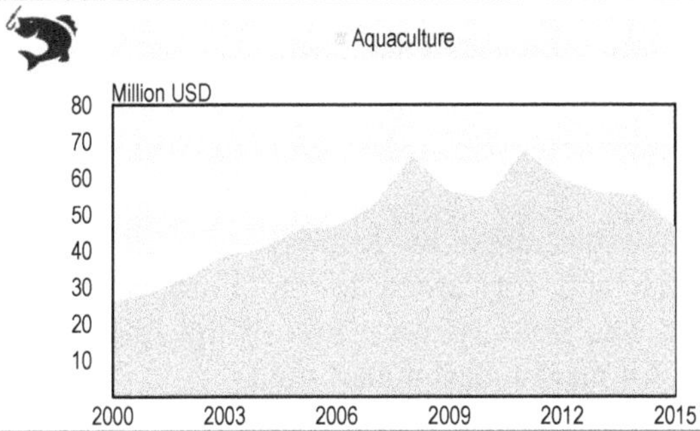

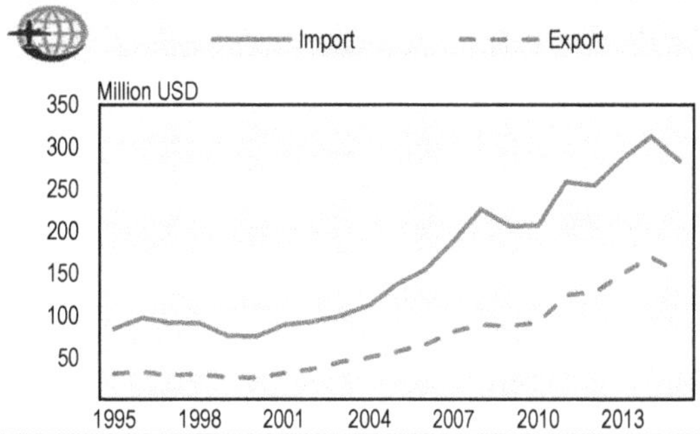

Employment and fleet

Employment in aquaculture increased by 5.1% between 2014 and 2015 to 1 506 persons

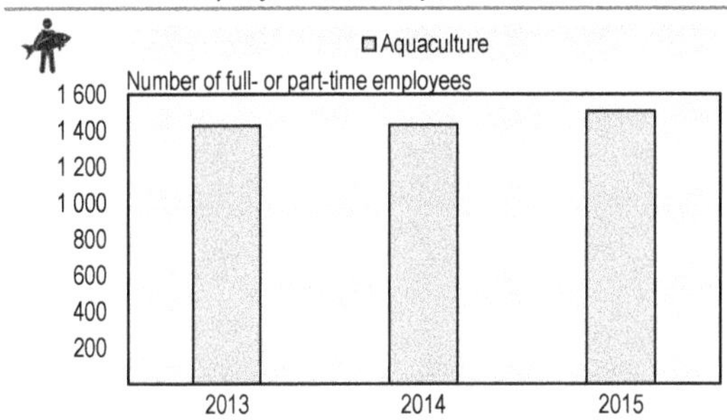

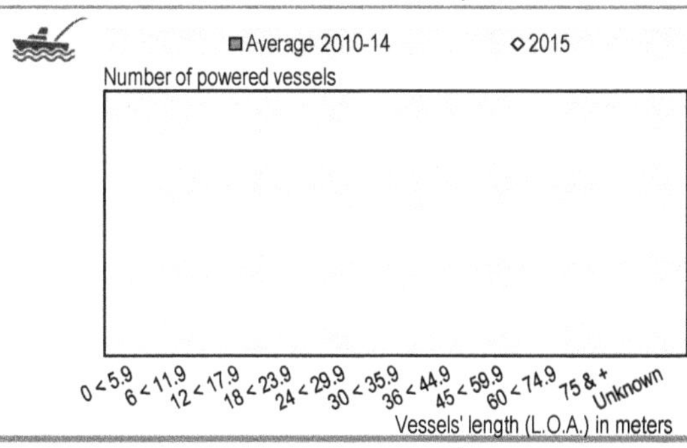

Government support and resource management

There is no marine fisheries production in the country

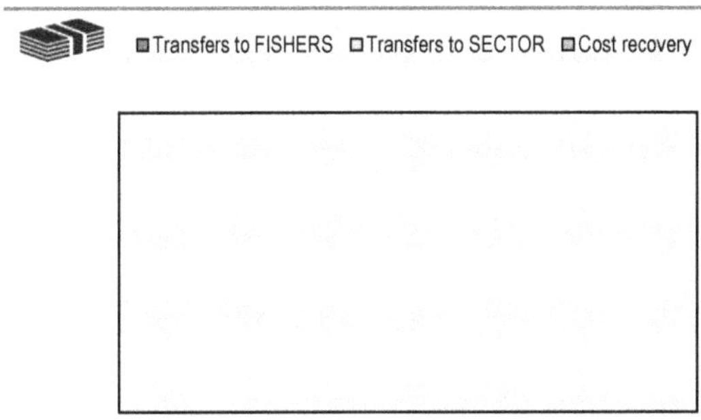

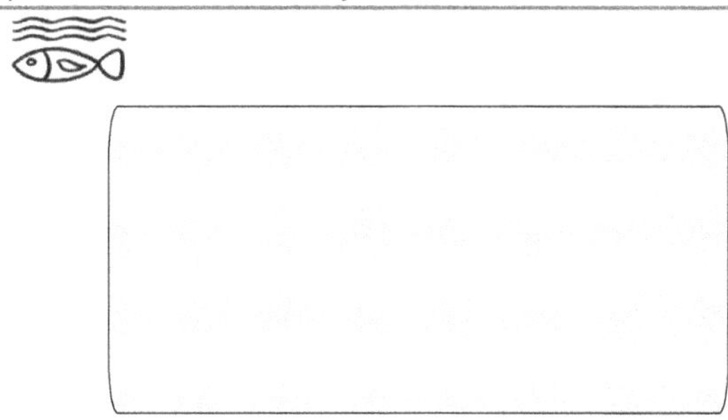

Note: Reference year is 2015.

Source: OECD Fisheries Database. World's total values of production for catch and aquaculture were sourced from the *FAO Yearbook, Fishery and Aquaculture Statistics, 2015*. Both numbers have been corrected adding the value of seaweed production; aquaculture has also been adjusted to account for the value of Chinese production reported to the OECD.

Denmark

Developments in fisheries and aquaculture policy, 2015-2016

- Denmark is preparing for the launch of a new scheme for coastal fisheries in 2017. More quotas for cod, sole and plaice will be transferred from the part of the quota reserved for vessels with individual transferable quotas (ITQs) to strengthen the sector. For the first time, quotas for Norway lobster, turbot, sprat, herring and saithe will be introduced. This initiative is part of an effort to improve the functioning of the transferable fishing concessions system in Denmark.
- The National Strategic Plan for Development of Sustainable Aquaculture 2014-2020 is in progress. The goal is a 25% increase in aquaculture output with a focus on eco-friendly production.
- New measures for the protection of reef structures in Kattegat and the Western Baltic Sea were formulated in 2016 and forwarded to the European Commission (EC) as joint recommendations. The initiatives are part of the final implementation of the Natura 2000 directives for marine designated sites.

Fisheries and aquaculture statistics in Denmark

Value of production

in global context

USD 629 million	0.4%	0.06%
Total value of fisheries and aquaculture	*Share of global fisheries landed value*	*Share of global aquaculture production value*

Trends in production value and trade

Total production value in USD decreased at an annual average rate of 4.7% between 2011 and 2015

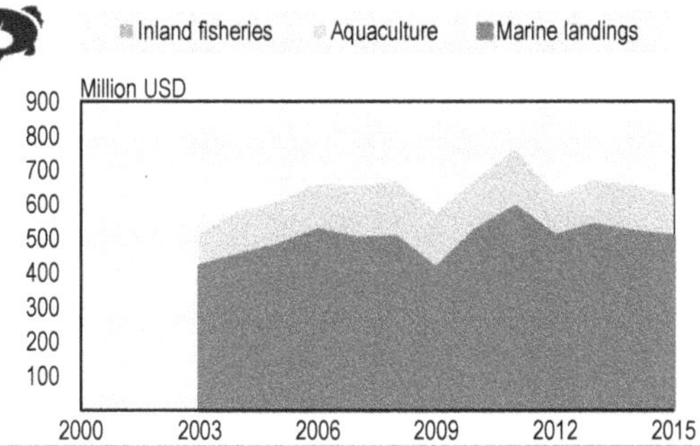

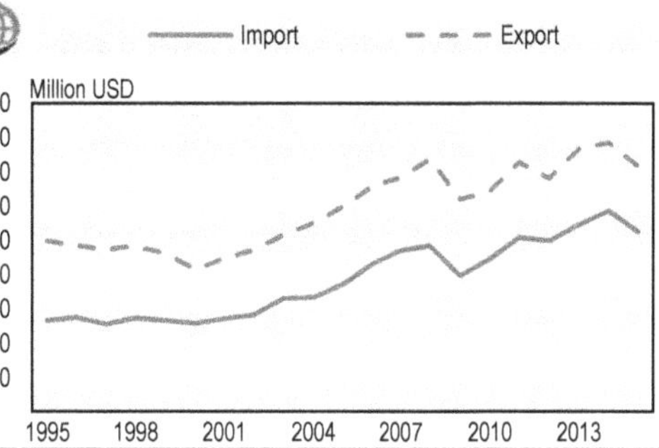

Employment and fleet

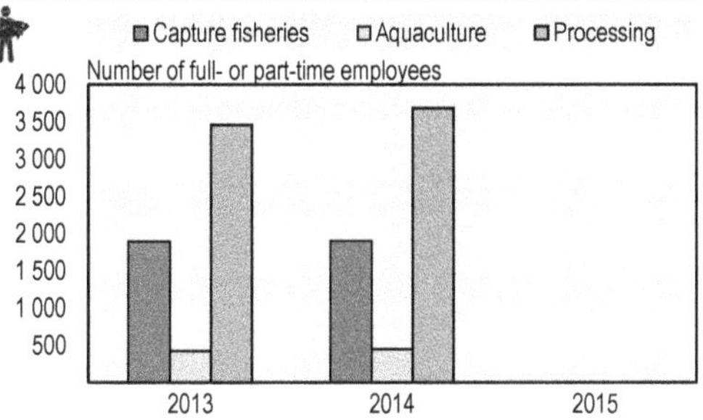

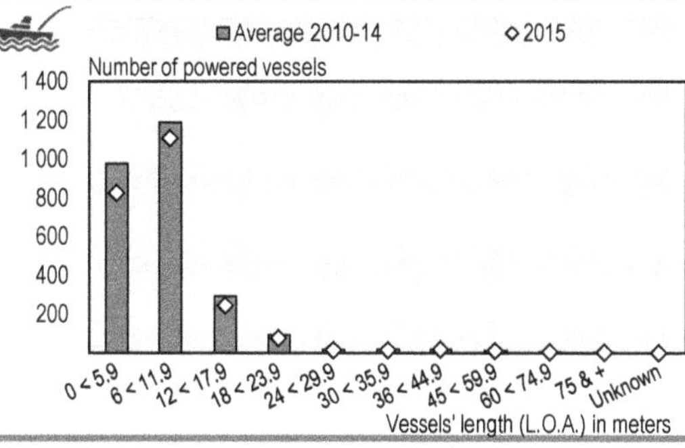

Government support and resource management

49% of total budgetary support was directed to management of resources in 2015

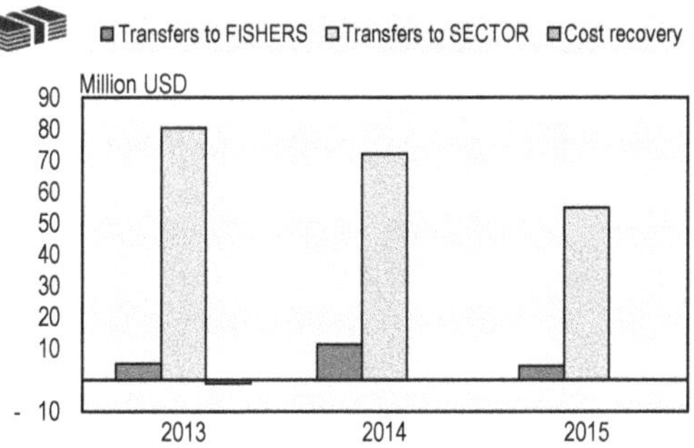

For information on EU fish stocks please refer to the EU country chapter

Note: Reference year is 2015.

Source: OECD Fisheries Database. World's total values of production for catch and aquaculture were sourced from the *FAO Yearbook, Fishery and Aquaculture Statistics, 2015*. Both numbers have been corrected adding the value of seaweed production; aquaculture has also been adjusted to account for the value of Chinese production reported to the OECD.

Estonia

Developments in fisheries and aquaculture policy, 2015-16

- A new smartphone application for fishers will improve and simplify electronic record-keeping of catches.
- A multi-annual plan to manage Baltic fish stocks is now in effect, covering Baltic herring, sprat, cod and salmon.
- Primary production remains stable although a Russian Federation embargo on European Union (EU) food imports, in place since 2014 and expanded in 2015, has affected prices and trade flows.

Fisheries and aquaculture statistics in Estonia

Value of production

in global context

USD 66 million	0.05%	0.002%
Total value of fisheries and aquaculture	*Share of global fisheries landed value*	*Share of global aquaculture production value*

Trends in production value and trade

Total production value in USD decreased at an annual average rate of 6.7% between 2011 and 2015

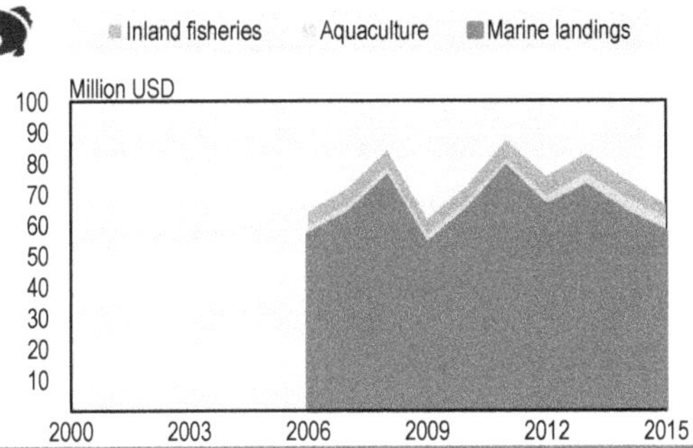

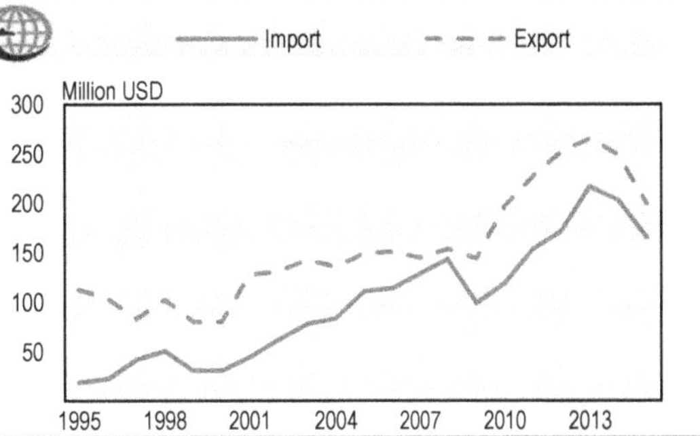

Employment and fleet

Employment in capture fisheries decreased by 1% between 2014 and 2015 to 2 884 persons

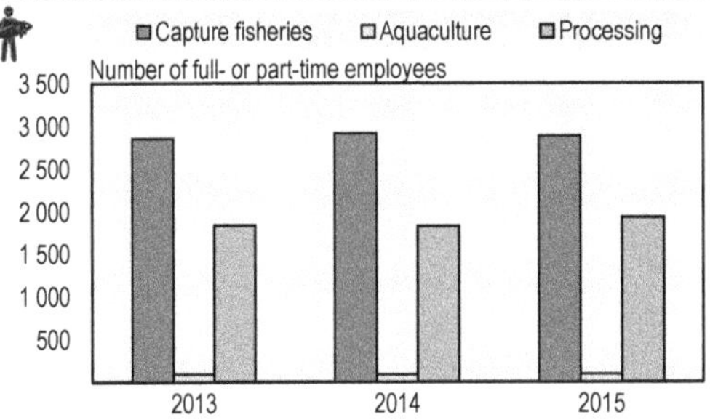

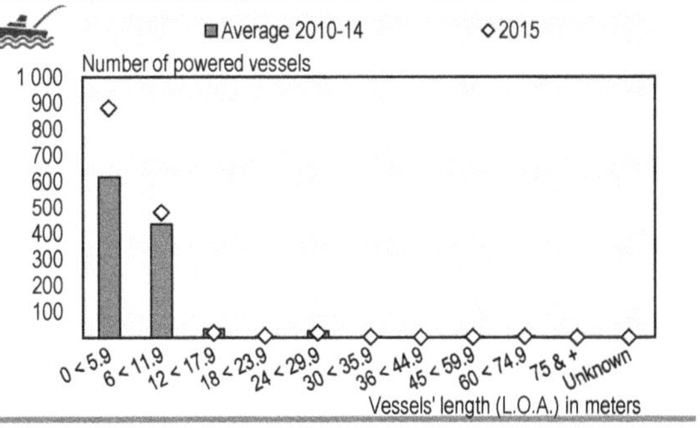

Government support and resource management

50% of total budgetary support was directed to management of resources in 2015

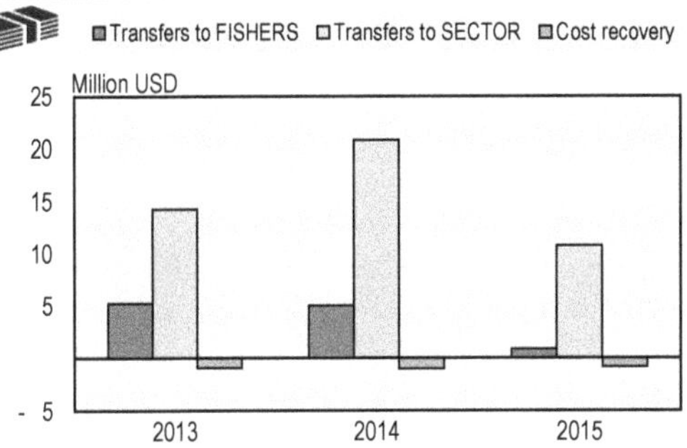

For information on EU fish stocks please refer to the EU country chapter

Note: Reference year is 2015. The 'Transfers to SECTOR' numbers reported here include estimations for management and enforcement expenditures, where missing.
Source: OECD Fisheries Database. World's total values of production for catch and aquaculture were sourced from the *FAO Yearbook, Fishery and Aquaculture Statistics, 2015*. Both numbers have been corrected adding the value of seaweed production; aquaculture has also been adjusted to account for the value of Chinese production reported to the OECD.

France

Developments in fisheries and aquaculture policy, 2015-16

- France has substantially increased the coverage of marine protected areas (MPAs) with almost 22% of waters under French jurisdiction designated as protected areas at the beginning of 2017, compared to 4% at the beginning of 2014.
- France is one of the largest aquaculture producers in the European Union (EU). In 2014, total aquaculture production amounted to 203 000 tonnes, of which 155 000 tonnes of farmed shellfish. 3 200 enterprises, employing over 10 000 people, are active in this sector, generating an overall turnover of EUR 730 million.
- In 2016, France embarked on a wide-ranging process of procedural harmonisation and simplification for enterprises in the fisheries sector.
- 61% of budgetary support was directed to management of resources in 2015.

Fisheries and aquaculture statistics in France

Value of production

in global context

---	---	---
Total value of fisheries and aquaculture	*Share of global fisheries landed value*	*Share of global aquaculture production value*

Trends in production value and trade

Value of net exports in USD decreased at an annual average rate of 3.8% between 2011 and 2015

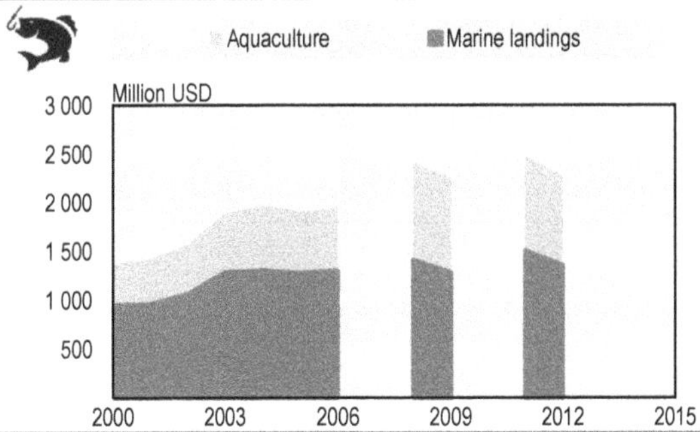

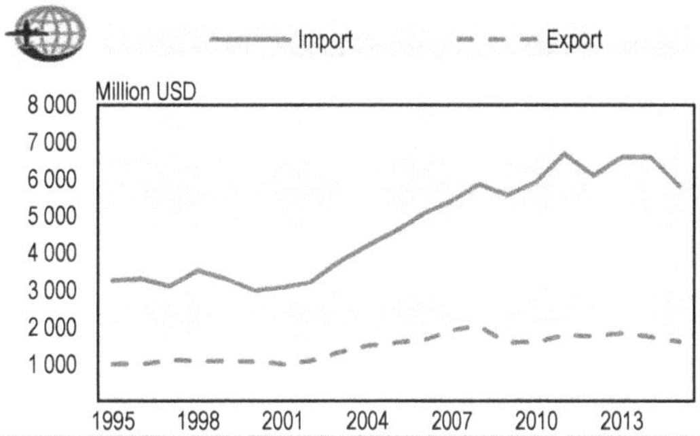

Employment and fleet

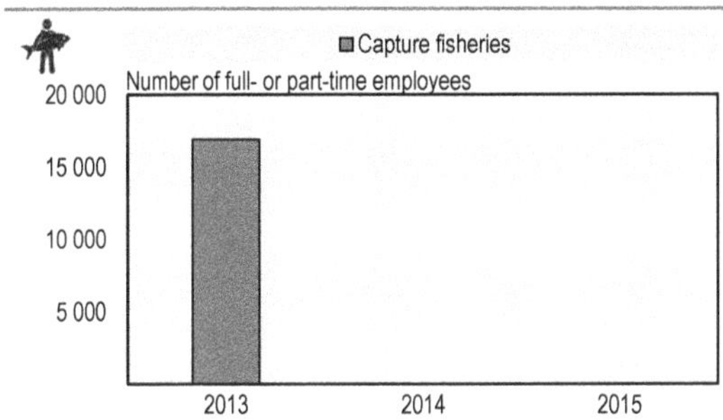

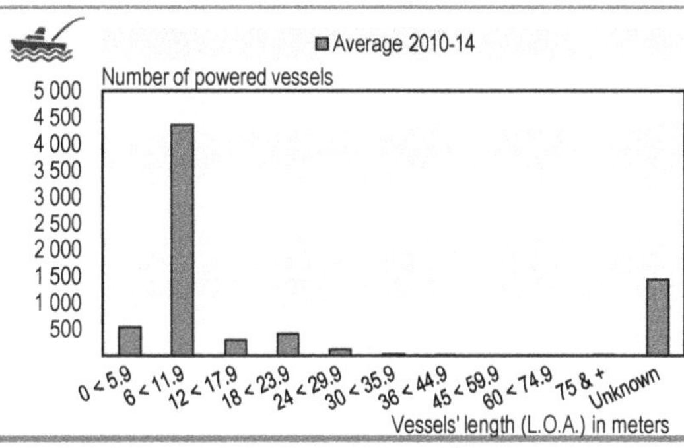

Government support and resource management

61% of total budgetary support was directed to management of resources in 2015

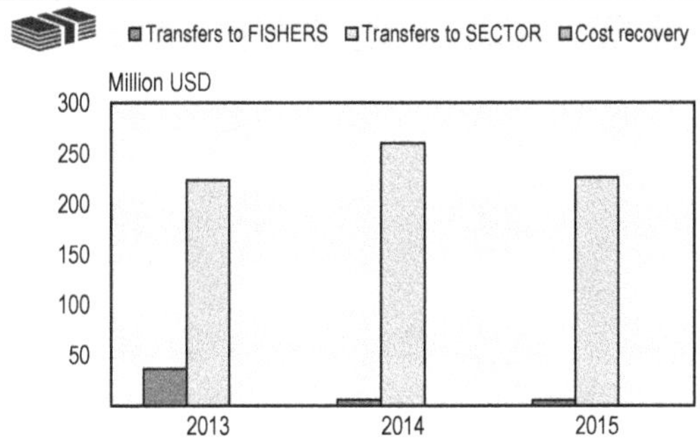

For information on EU fish stocks please refer to the EU country chapter

Note: Reference year is 2015. The 'Transfers to SECTOR' numbers reported here include estimations for management and enforcement expenditures, where missing.

Source: OECD Fisheries Database. World's total values of production for catch and aquaculture were sourced from the *FAO Yearbook, Fishery and Aquaculture Statistics, 2015*. Both numbers have been corrected adding the value of seaweed production; aquaculture has also been adjusted to account for the value of Chinese production reported to the OECD.

Germany

Developments in fisheries and aquaculture policy, 2015-16

- A 2014 notification allowing enterprises to improve efficiency and reduce capacity without the risk of consequently being allocated less quota resulted in five successfully completed proceedings between 2015 and 2016.
- Decommissioning has resulted in the continued reduction of German fishing vessel numbers, predominantly those less than 12 m in length, but the fleet's overall tonnage has increased.
- Changes to Germany's Sea Fisheries Act in 2015 have increased the country's capacity to penalise infringements of fisheries rules.

Fisheries and aquaculture statistics in Germany

Value of production

in global context

USD 312 million	0.2%	0.05%
Total value of fisheries and aquaculture	*Share of global fisheries landed value*	*Share of global aquaculture production value*

Trends in production value and trade

Total production value in USD decreased at an annual average rate of 5.4% between 2011 and 2015

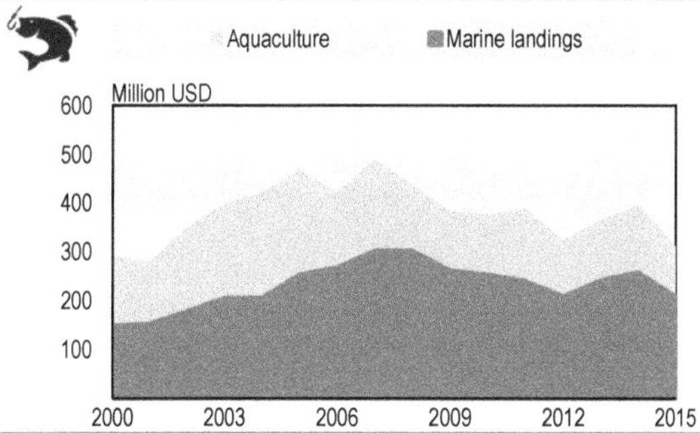

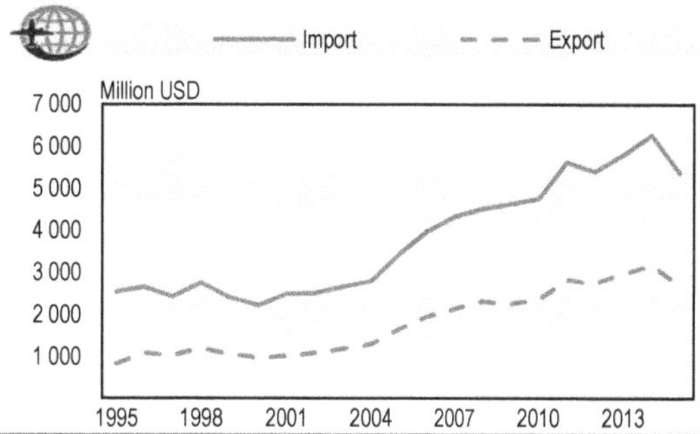

Employment and fleet

Employment in capture fisheries decreased by 0.5% between 2014 and 2015 to 4 404 persons

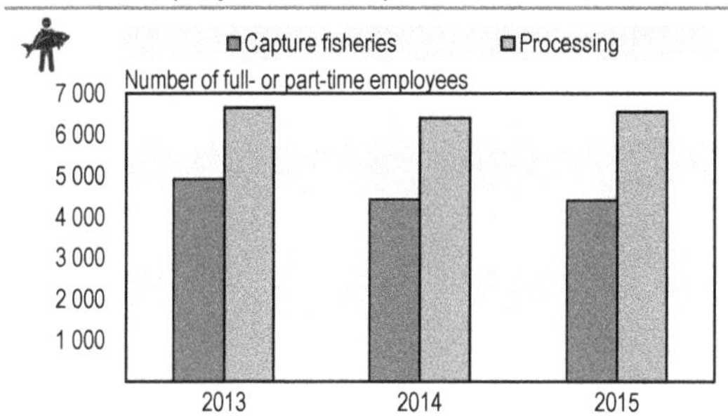

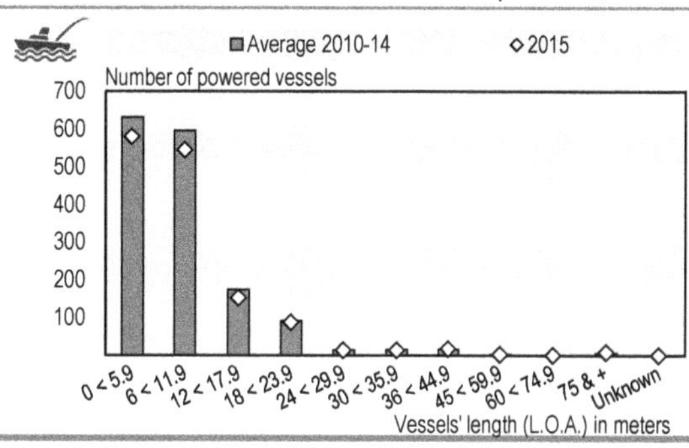

Government support and resource management

44% of total budgetary support was directed to management of resources in 2015

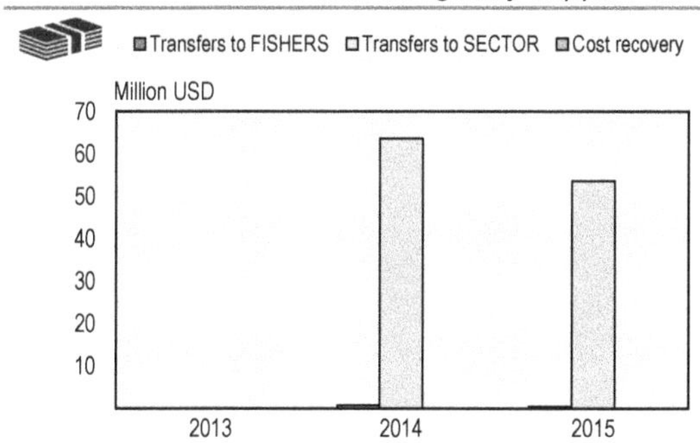

For information on EU fish stocks please refer to the EU country chapter

Note: Reference year is 2015. The 'Transfers to SECTOR' numbers reported here include estimations for management and enforcement expenditures, where missing.

Source: OECD Fisheries Database. World's total values of production for catch and aquaculture were sourced from the *FAO Yearbook, Fishery and Aquaculture Statistics, 2015*. Both numbers have been corrected adding the value of seaweed production; aquaculture has also been adjusted to account for the value of Chinese production reported to the OECD.

Greece

Developments in fisheries and aquaculture policy, 2015-16

- National planning for aquaculture and simplified formalities for licensing have been put in place to encourage growth in the sector.
- Profits have increased as fuel costs decline, but the long-term economic viability of the small-scale fleet remains uncertain.
- Gear bans have been established in certain regions following a seagrass mapping exercise that identified vulnerable areas.

Fisheries and aquaculture statistics in Greece

Value of production

in global context

USD 804 million	0.2%	0.3%
Total value of fisheries and aquaculture	*Share of global fisheries landed value*	*Share of global aquaculture production value*

Trends in production value and trade

Total production value in USD decreased at an annual average rate of 4.9% between 2011 and 2015

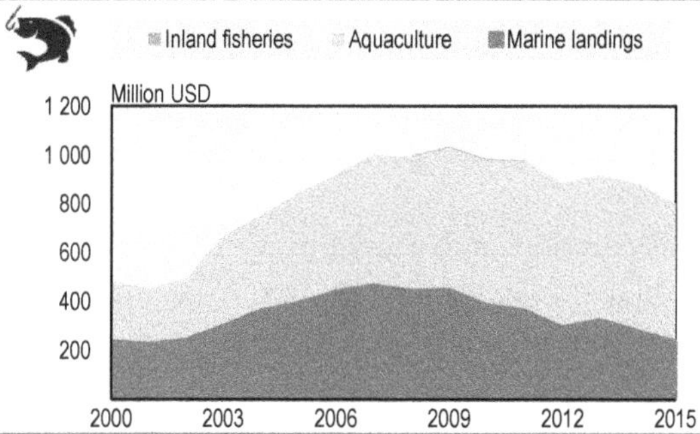

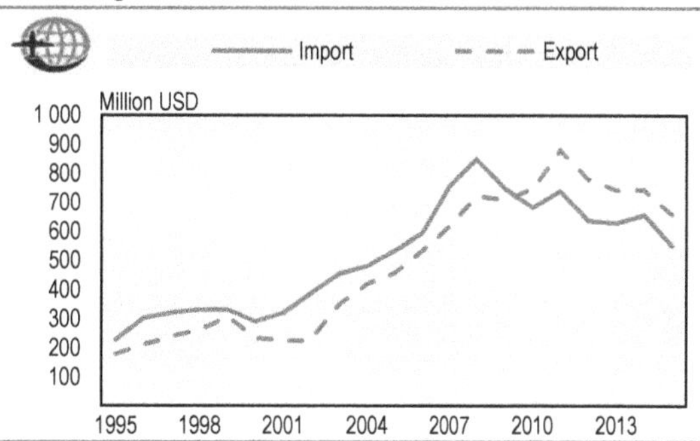

Employment and fleet

Employment in capture fisheries decreased by 2.4% between 2014 and 2015 to 34 770 persons

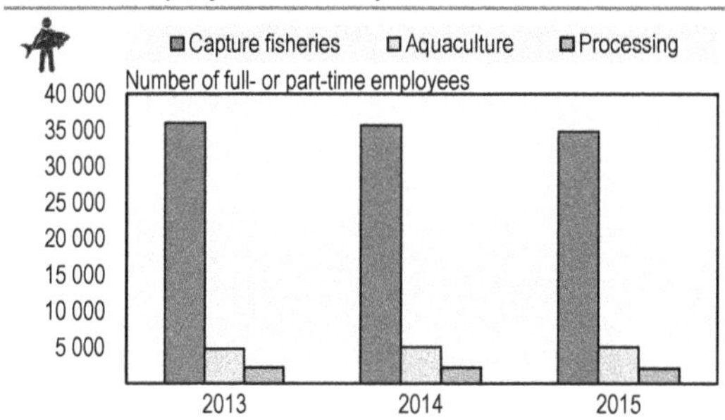

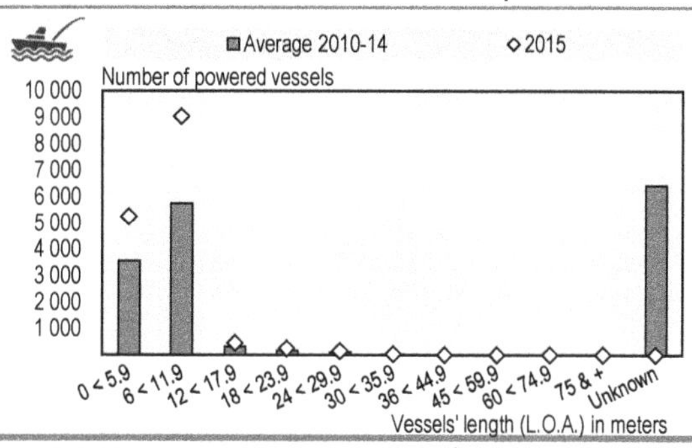

Government support and resource management

43% of total budgetary support was directed to management of resources in 2015

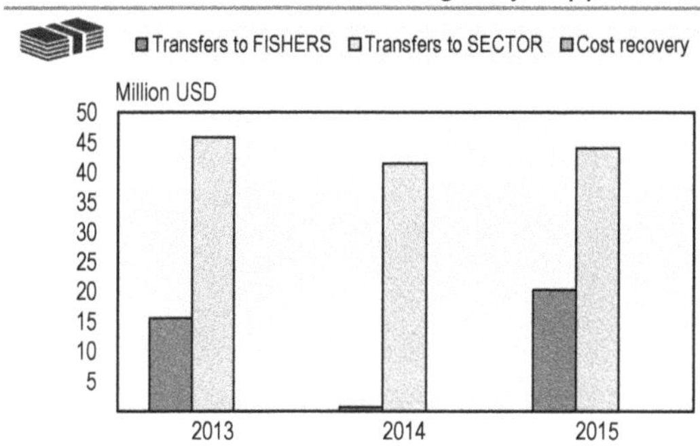

For information on EU fish stocks please refer to the EU country chapter

Note: Reference year is 2015. The 'Transfers to SECTOR' numbers reported here include estimations for management and enforcement expenditures, where missing.

Source: OECD Fisheries Database. World's total values of production for catch and aquaculture were sourced from the *FAO Yearbook, Fishery and Aquaculture Statistics, 2015*. Both numbers have been corrected adding the value of seaweed production; aquaculture has also been adjusted to account for the value of Chinese production reported to the OECD.

Iceland

Developments in fisheries and aquaculture policy, 2015-16

- Identifying an approach to ensure the resource rent tax that vessel operators pay on landed catch reflects the real economic rents accrued by the industry is an ongoing policy priority.
- There is a push to increase aquaculture output and permitted levels of production increased in 2015. Production has been following an increasing trend in recent years and this is expected to continue in 2016-17.
- Iceland is an export-oriented seafood producer and the appreciating krona, driven by recent increases in tourism, has put downward pressure on the value of export earnings.
- A changing climate is influencing the distribution and accessibility of some stocks, necessitating continued international collaboration on the issue and requiring future policy decisions to account for these changes.

Fisheries and aquaculture statistics in Iceland

Value of production

in global context

USD 1 117 million	0.9%	----
Total value of fisheries and aquaculture	*Share of global fisheries landed value*	*Share of global aquaculture production value*

Trends in production value and trade

Total production value in USD decreased at an annual average rate of 2.3% between 2011 and 2015

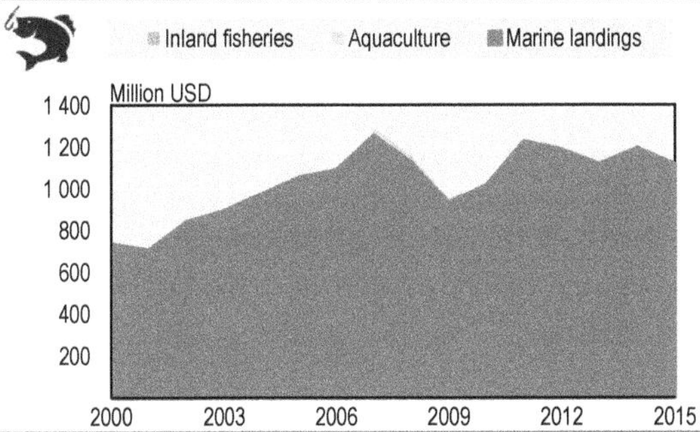

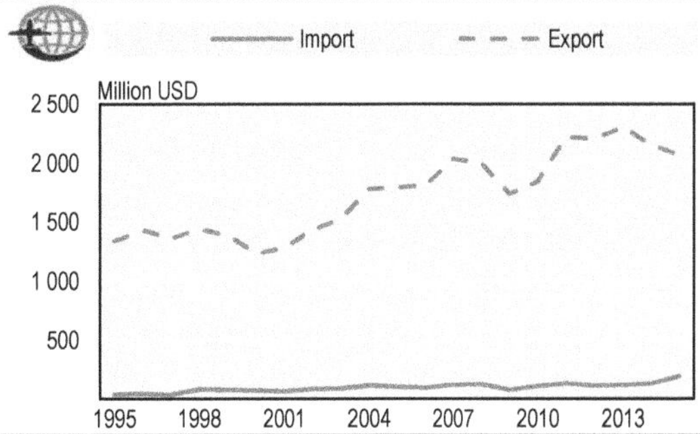

Employment and fleet

Employment in capture fisheries decreased by 9.1% between 2014 and 2015 to 4 000 persons

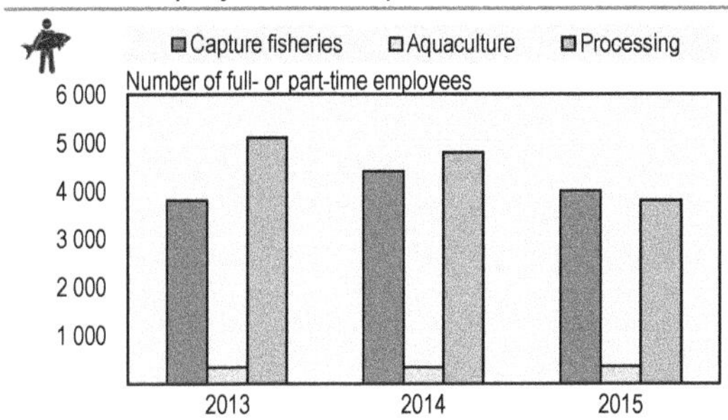

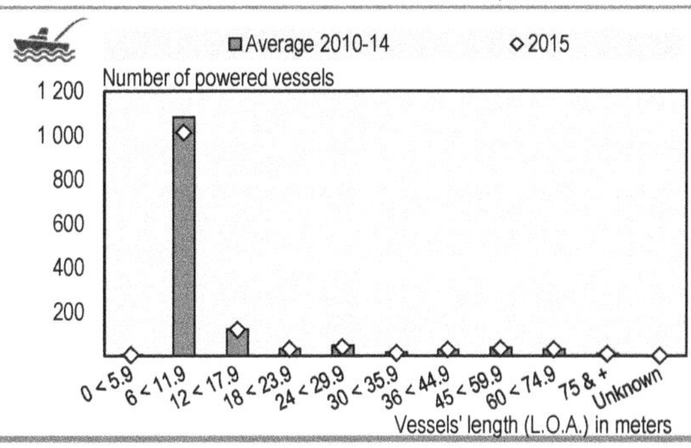

Government support and resource management

Cost recovery is equal to 81% of total support to the sector in 2015

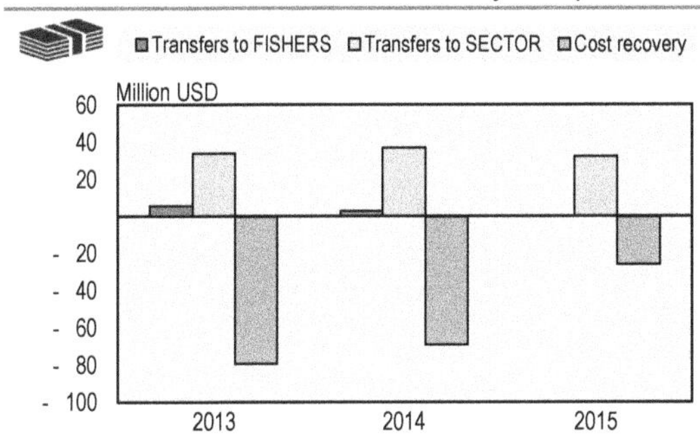

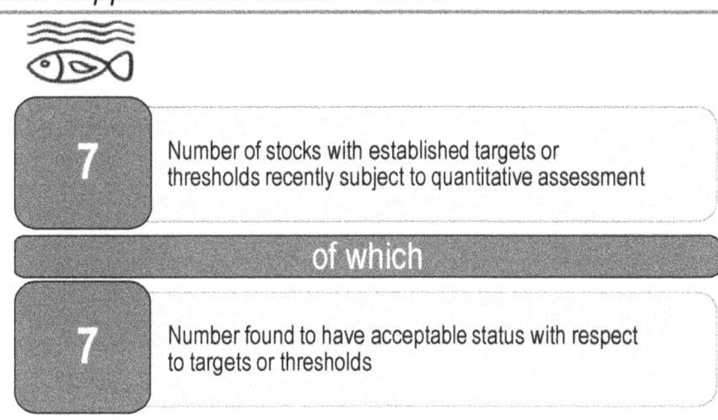

Note: Reference year is 2015, except for stock assessment data which may cover more than one year.

Source: OECD Fisheries Database. World's total values of production for catch and aquaculture were sourced from the *FAO Yearbook, Fishery and Aquaculture Statistics, 2015*. Both numbers have been corrected adding the value of seaweed production; aquaculture has also been adjusted to account for the value of Chinese production reported to the OECD. Fish stocks data were directly reported by the country.

Indonesia

Developments in fisheries and aquaculture policy, 2015-16

- Indonesia has become the second-largest producer of fish, crustaceans and aquatic plants globally by volume, after the People's Republic of China (hereafter "China"), notably due to a boom in aquaculture production dominated by seaweed.
- Since the election of President Widodo in 2014, maritime and fisheries policy has been a central priority for the Indonesian government. The budget of the Ministry of Marine Affairs and Fisheries (MMAF) doubled in 2015.
- The Indonesian government has taken significant measures to combat illegal fishing piloted by the Presidential Task Force to Combat Illegal Fishing, which co-ordinates the actions of all administrative bodies concerned under the authority of the Minister of Marine Affairs and Fisheries. A permanent moratorium on fishing by ex-foreign vessels operating within the exclusive economic zone (EEZ) and a ban on transhipment at sea were adopted in 2014.

Fisheries and aquaculture statistics in Indonesia

Value of production

in global context

USD 18 073 million	7.4%	4.2%
Total value of fisheries and aquaculture	*Share of global fisheries landed value*	*Share of global aquaculture production value*

Trends in production value and trade

Total production value in USD increased at an annual average rate of 4.3% between 2011 and 2015

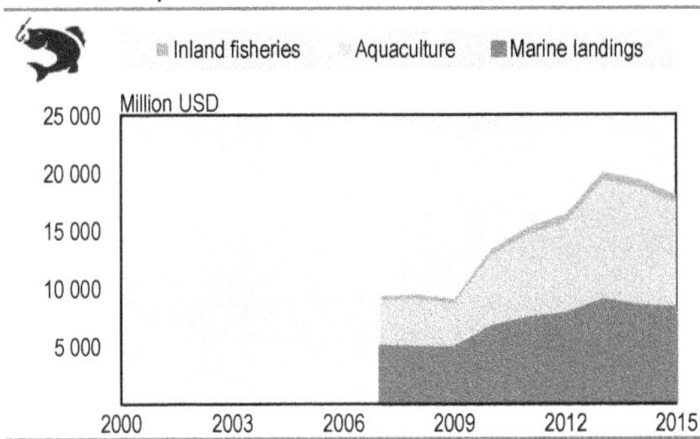

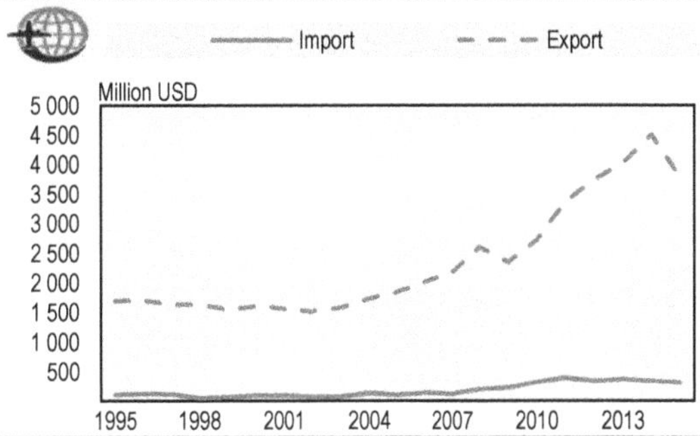

Employment and fleet

Employment in capture fisheries decreased by 1.4% between 2014 and 2015 to 2 702 664 persons

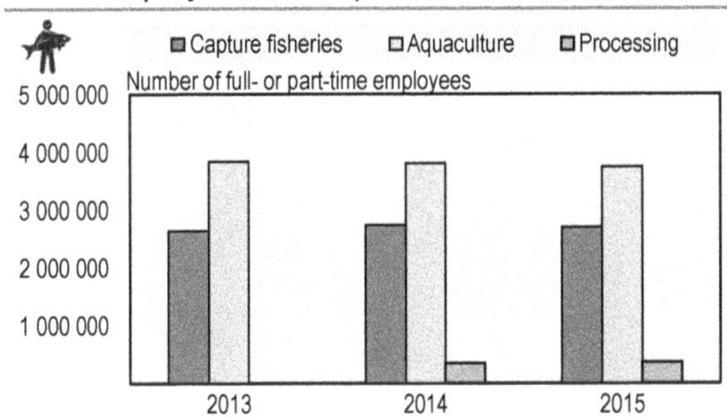

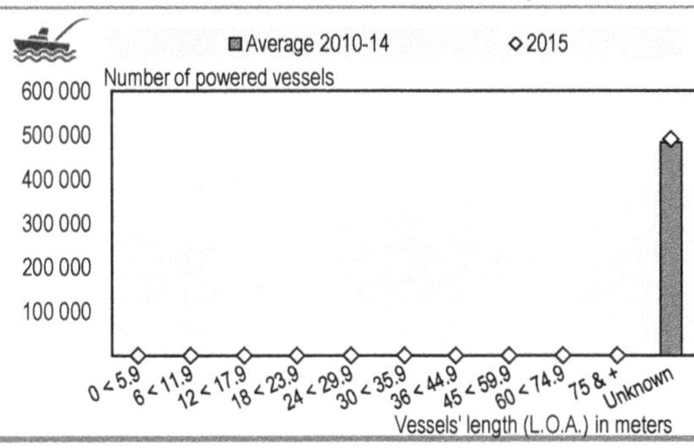

Government support and resource management

55% of total budgetary support was directed to management of resources in 2015

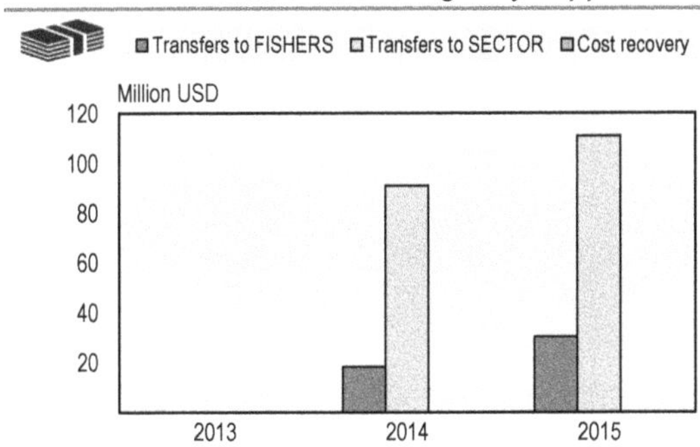

8 Number of stocks with established targets or thresholds recently subject to quantitative assessment

of which

4 Number found to have acceptable status with respect to targets or thresholds

Note: Reference year is 2015, except for stock assessment data which may cover more than one year. The 'Transfers to SECTOR' numbers reported here include estimations for management and enforcement expenditures, where missing.

Source: OECD Fisheries Database. World's total values of production for catch and aquaculture were sourced from the *FAO Yearbook, Fishery and Aquaculture Statistics, 2015*. Both numbers have been corrected adding the value of seaweed production; aquaculture has also been adjusted to account for the value of Chinese production reported to the OECD. Fish stocks data were directly reported by the country.

Ireland

Developments in fisheries and aquaculture policy, 2015-16

- The National Inshore Fisheries Forum (NIFF), established by the Irish Government to facilitate the discussion of inshore fisheries issues and to foster industry-led development of proposals, is now operational and met nine times over the 2015-16 period.
- In its 2015 National Strategic Plan for Sustainable Aquaculture Development the Irish Government introduced a set of 'Guiding Principles for the Sustainable Development of Aquaculture', and 'Scale Limits for Individual Offshore Salmon Farms'.
- A review of Ireland's regulatory framework for aquaculture licensing was initiated in 2016 with recommendations due in the first half of 2017.
- By the end of 2016 the assessment of likely interactions between aquaculture activities and conservation features of designated NATURA sites had been completed for 22 bays with aquaculture activities in Ireland.

Fisheries and aquaculture statistics in Ireland

Value of production

in global context

USD 421 million	0.2%	0.07%
Total value of fisheries and aquaculture	*Share of global fisheries landed value*	*Share of global aquaculture production value*

Trends in production value and trade

Total production value in USD decreased at an annual average rate of 4.8% between 2011 and 2015

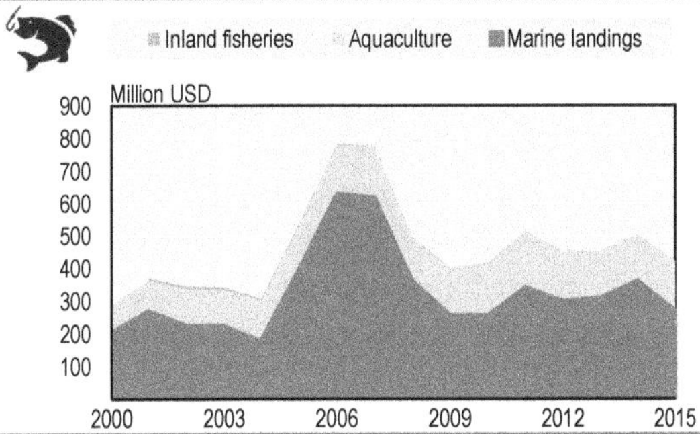

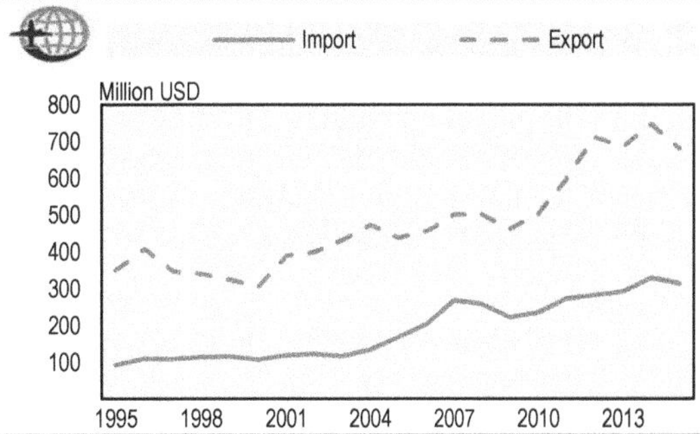

Employment and fleet

Employment in capture fisheries increased by 9.4% between 2014 and 2015 to 3 451 persons

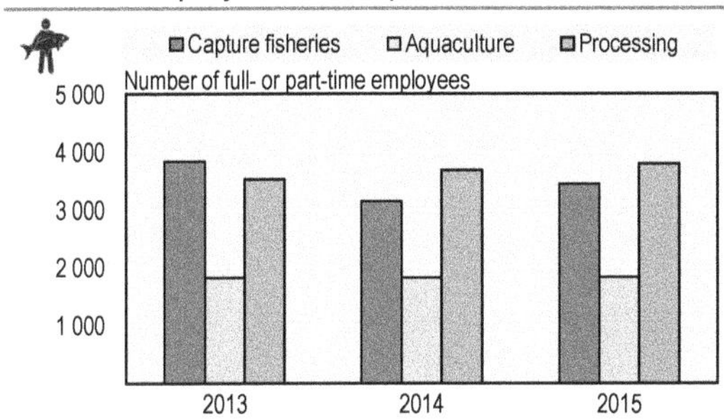

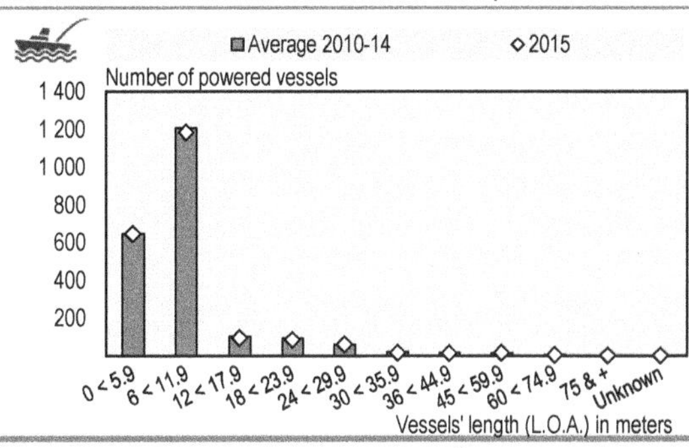

Government support and resource management

58% of total budgetary support was directed to management of resources in 2015

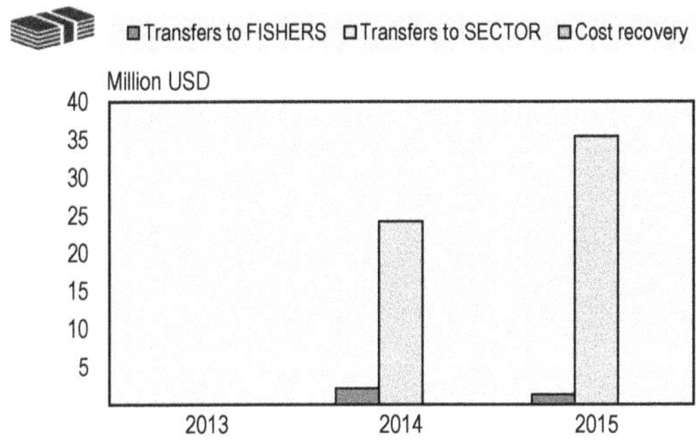

For information on EU fish stocks please refer to the EU country chapter

Note: Reference year is 2015.

Source: OECD Fisheries Database. World's total values of production for catch and aquaculture were sourced from the *FAO Yearbook, Fishery and Aquaculture Statistics, 2015*. Both numbers have been corrected adding the value of seaweed production; aquaculture has also been adjusted to account for the value of Chinese production reported to the OECD.

Italy

Developments in fisheries and aquaculture policy, 2015-16

- Italy is in the process of implementing the second phase of its National Operational Programme for fisheries and aquaculture. The key objective is to improve the competitiveness of its seafood industry. The total budget of the programme is nearly EUR 1 billion (USD 1.3 billion).
- The portfolio of regulations targeting small-scale fishers is growing. Italy is focusing on reversing the decline of coastal communities and allocates funds to support sustainable development of traditionally fisheries-dependent areas. It also promotes diversification into tourist services and recreational fishing.
- Achieving the United Nations Sustainable Development Goals (SDGs) for Good Environmental Status of the seas defined in the Marine Strategy Framework Directive is challenging in Italy. The measures adopted through the regulatory instruments of the fisheries and aquaculture sectors are intended to serve this purpose.

Fisheries and aquaculture statistics in Italy

Value of production

in global context

USD 1 446 million	0.8%	0.2%
Total value of fisheries and aquaculture	*Share of global fisheries landed value*	*Share of global aquaculture production value*

Trends in production value and trade

Total production value in USD decreased at an annual average rate of 9.6% between 2011 and 2015

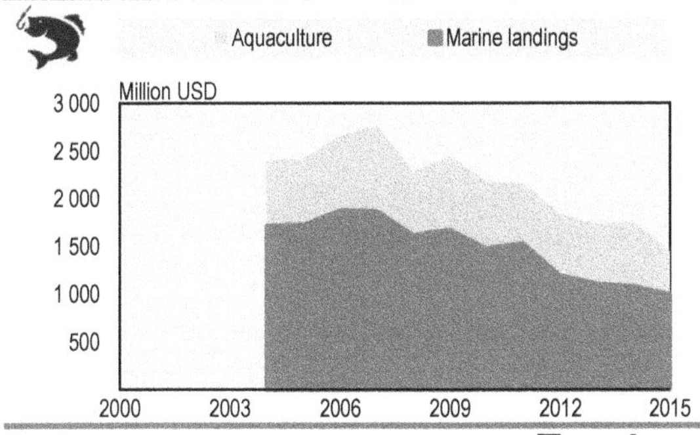

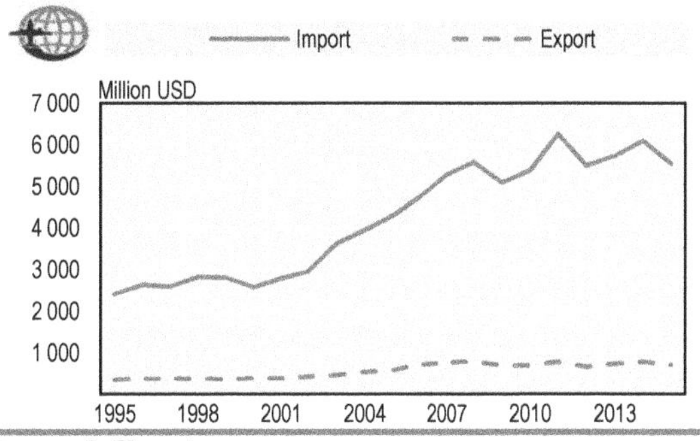

Employment and fleet

Employment in capture fisheries decreased by 4.2% between 2014 and 2015 to 25 811 persons

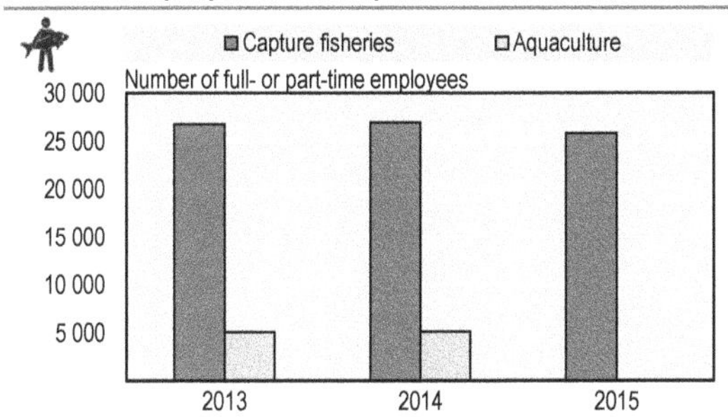

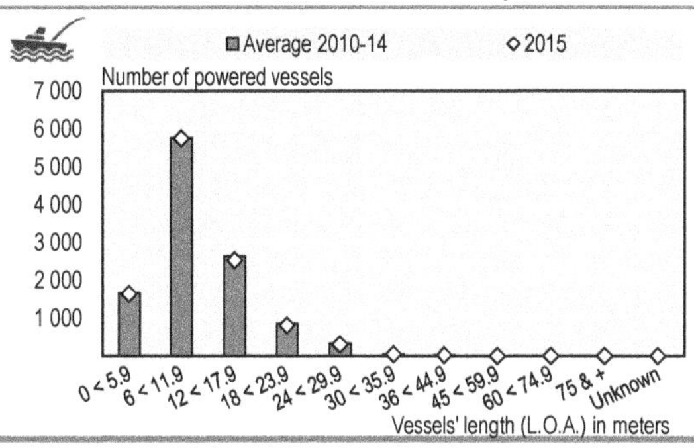

Government support and resource management

49% of total budgetary support was directed to management of resources in 2015

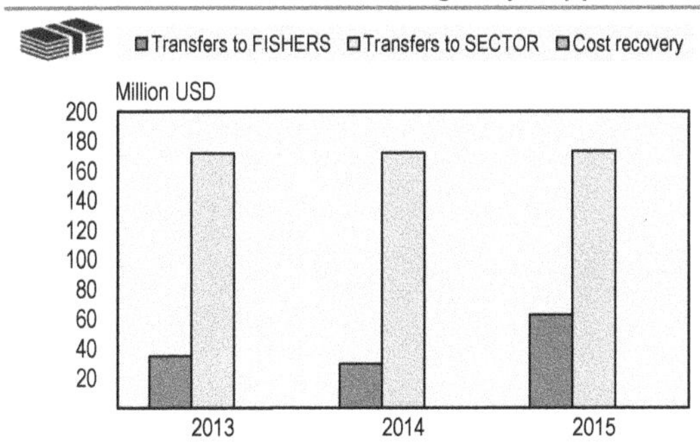

For information on EU fish stocks please refer to the EU country chapter

Note: Reference year is 2015. The 'Transfers to SECTOR' numbers reported here include estimations for management and enforcement expenditures, where missing.

Source: OECD Fisheries Database. World's total values of production for catch and aquaculture were sourced from the *FAO Yearbook, Fishery and Aquaculture Statistics, 2015*. Both numbers have been corrected adding the value of seaweed production; aquaculture has also been adjusted to account for the value of Chinese production reported to the OECD.

Japan

Developments in fisheries and aquaculture policy, 2015-16

- Government support to the fisheries sector in 2015 was JPY 227 billion (USD 1.9 billion), representing an increase of 3% from 2014. Most of the support went to general services for the fisheries sector such as management resources and infrastructure.
- In 2015, the Japanese government set catch quotas for Pacific bluefin tuna for troll fishing, set-net fishing and purse seine fishery in accordance with the Conservation and Management Measure adopted by the Western and Central Pacific Fisheries Commission (WCPFC).
- The 7th Basic Plan for Fish Farming in Japan was established in 2015. The plan aims to further promote fish farming for aquaculture production and to restore depleted fish stocks.
- In 2016, the Fisheries Agency developed the Seaweed Beds and Tidal Flats Vision to conserve these areas and restore habitats for fisheries resources. The Vision proposes practical measures in response to the shrinkage of these areas.

Fisheries and aquaculture statistics in Japan

Value of production

in global context

USD 12 991 million	7.0%	2.1%
Total value of fisheries and aquaculture	*Share of global fisheries landed value*	*Share of global aquaculture production value*

Trends in production value and trade

Total production value in USD decreased at an annual average rate of 7.3% between 2011 and 2015

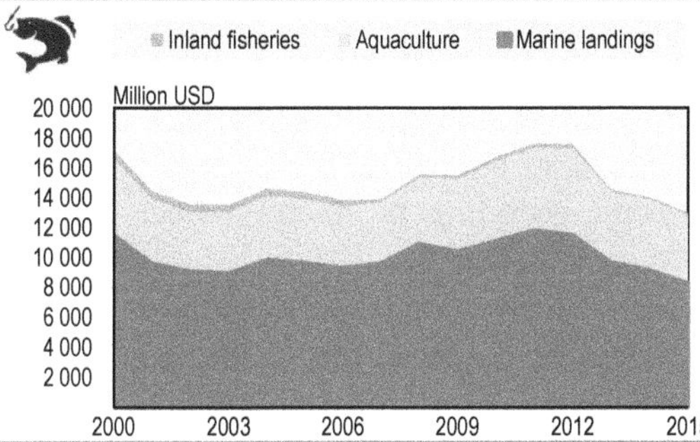

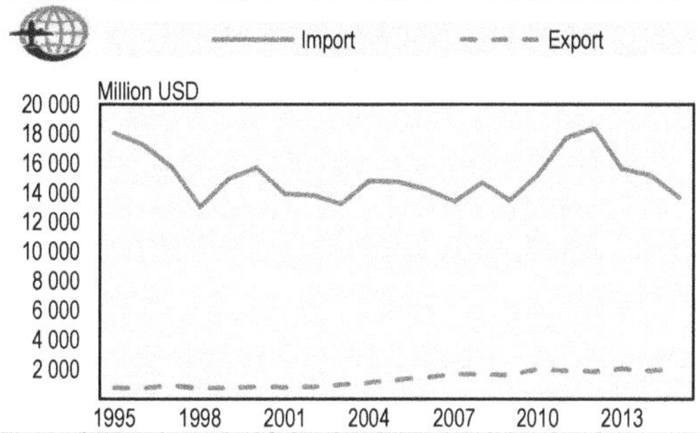

Employment and fleet

Employment in capture fisheries decreased by 3.7% between 2014 and 2015 to 166 610 persons

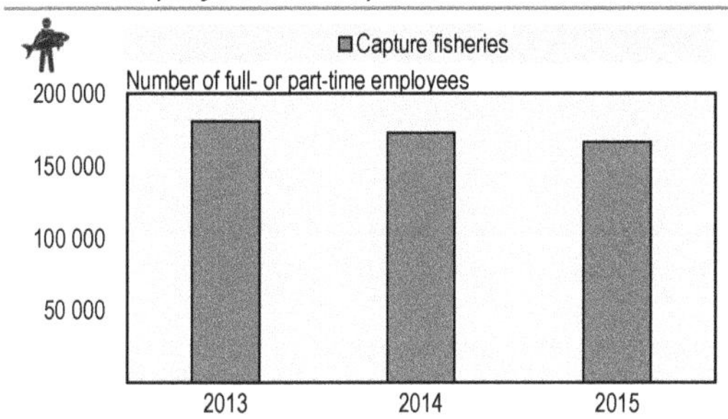

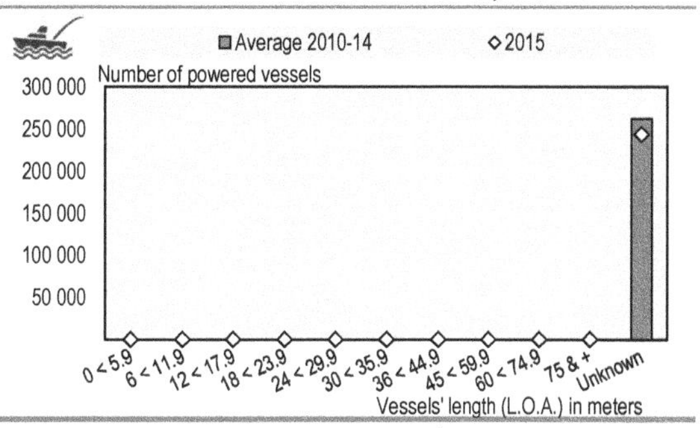

Government support and resource management

55% of total budgetary support was directed to management of resources in 2015

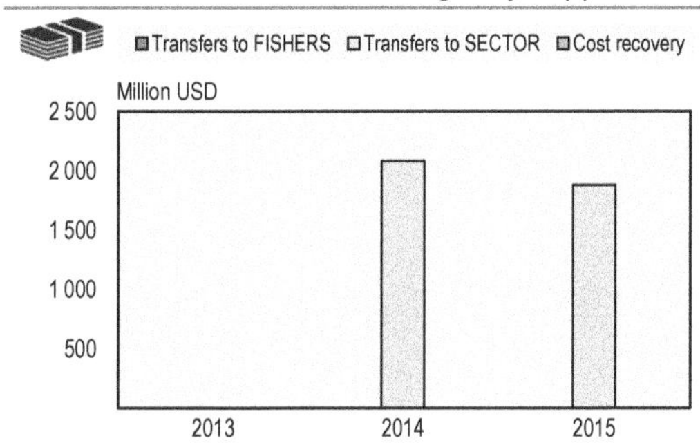

84 Number of stocks with established targets or thresholds recently subject to quantitative assessment

of which

43 Number where stock level was reported to be either medium or high

Note: Reference year is 2015, except for stock assessment data which may cover more than one year. The 'Transfers to SECTOR' numbers reported here include estimations for management and enforcement expenditures, where missing.

Source: OECD Fisheries Database. World's total values of production for catch and aquaculture were sourced from the *FAO Yearbook, Fishery and Aquaculture Statistics, 2015*. Both numbers have been corrected adding the value of seaweed production; aquaculture has also been adjusted to account for the value of Chinese production reported to the OECD. Fish stocks data were sourced from Japan's fish classification system group stock assessment. A stock is considered to be at a medium level when above Blim or the CPUE threshold and high when well above these.

Korea

Developments in fisheries and aquaculture policy, 2015-16

- In 2016, Korea adopted a five year master plan (2016-2020) for the development of fisheries related industries and fishing villages. The plan supports the sustainability of fisheries, food safety and fishing communities. The new integrated plan aims to increase policy effectiveness, by harmonising previously fragmented plans.
- The government's support to the fisheries sector in 2015 was KRW 482 billion (USD 426 million), a 29% decrease compared with the 2009-2011 average. Support policies for the sector continue to provide low interest loans to small fishers to secure their businesses, and vessel decommissioning to maintain fleet size at sustainable levels.
- In 2016, the vessel buyback programme underwent reforms. The government can now designate special vessel types which could deplete fisheries resources and require vessels demonstrating a history of illegal, unreported and unregulated (IUU) fishing to register for the programme starting in 2017.
- The government enacted the Fish Seed Industry Development Law in 2015 to develop the aquaculture industry and launched a pilot project to provide consultant services to fish seed businesses. An aircraft-mounted camera monitoring system, introduced in 2015, improved analysis and estimation of production and market prices.

Fisheries and aquaculture statistics in Korea

Value of production

in global context

USD 5 444 million	2.9%	0.9%
Total value of fisheries and aquaculture	*Share of global fisheries landed value*	*Share of global aquaculture production value*

Trends in production value and trade

Total production value in USD decreased at an annual average rate of 1.9% between 2011 and 2015

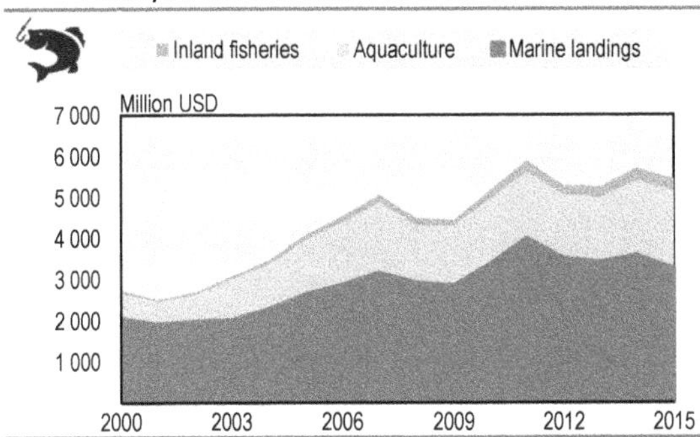

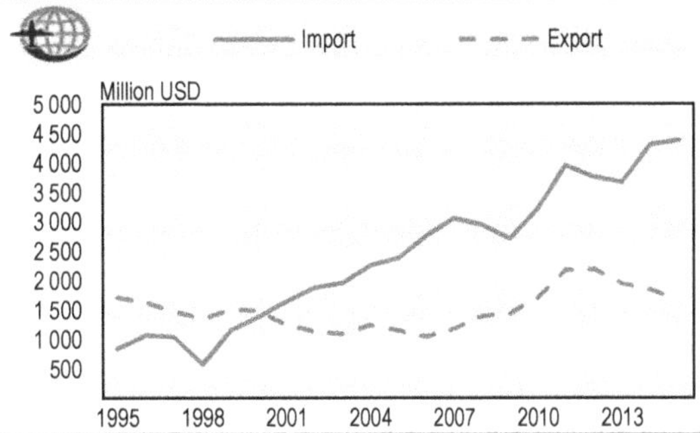

Employment and fleet

Total number of the fleet decreased by 8.5% in 2015 versus the average in 2010-2014

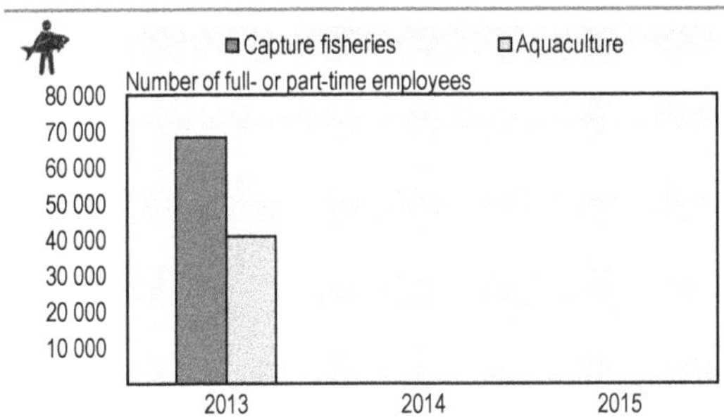

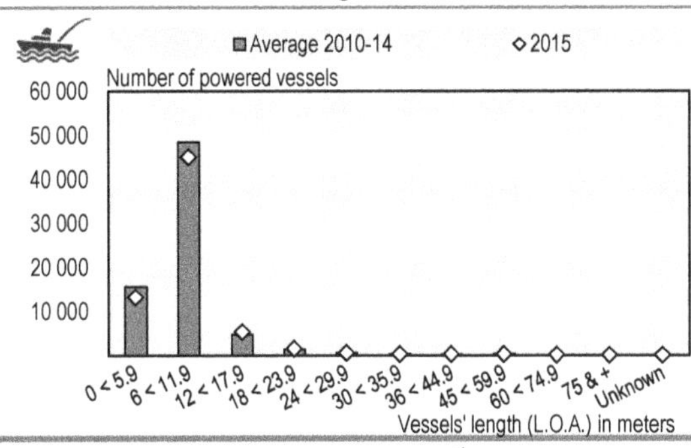

Government support and resource management

88% of total budgetary support was directed to management of resources in 2015

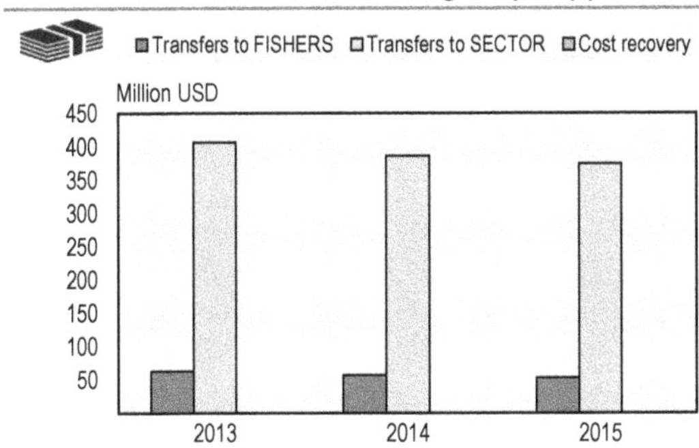

---	Number of stocks with established targets or thresholds recently subject to quantitative assessment
of which	
---	Number of stocks found to be within acceptable conditions

Note: Reference year is 2015. The 'Transfers to SECTOR' numbers reported here include estimations for management and enforcement expenditures, where missing.

Source: OECD Fisheries Database. World's total values of production for catch and aquaculture were sourced from the *FAO Yearbook, Fishery and Aquaculture Statistics, 2015*. Both numbers have been corrected adding the value of seaweed production; aquaculture has also been adjusted to account for the value of Chinese production reported to the OECD.

Latvia

Developments in fisheries and aquaculture policy, 2015-2016

- An electronic fish product traceability system is in development to improve control of fishing activity. The existing traceability system is based on paper documents with a unique traceability number. The new electronic system provides more flexibility, convenience and reduces paper use.
- New technical control measures for vessels participating in Baltic fisheries are under development within the context of the Baltic Sea Fisheries Forum (BALTFISH).

Fisheries and aquaculture statistics in Latvia

Value of production

in global context

USD 44 million	0.03%	0.001%
Total value of fisheries and aquaculture	*Share of global fisheries landed value*	*Share of global aquaculture production value*

Trends in production value and trade

Total production value in USD decreased at an annual average rate of 9.3% between 2011 and 2015

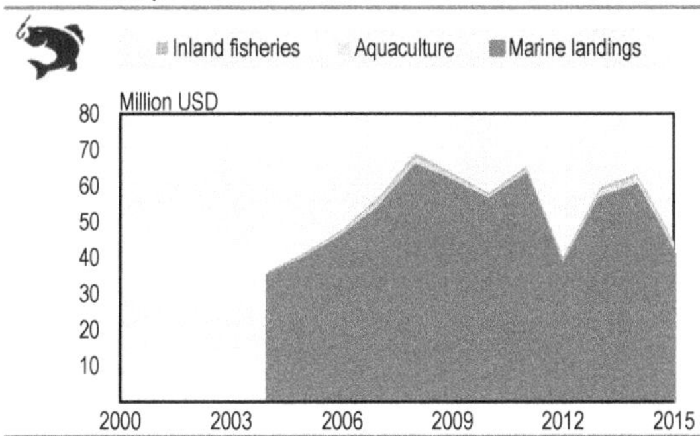

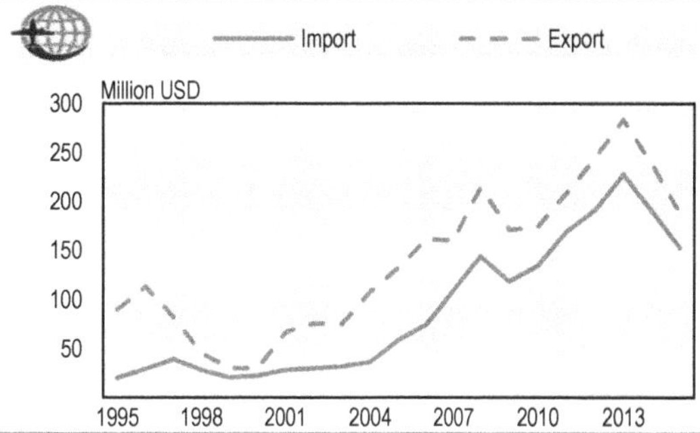

Employment and fleet

Employment in capture fisheries increased by 12.1% between 2014 and 2015 to 711 persons

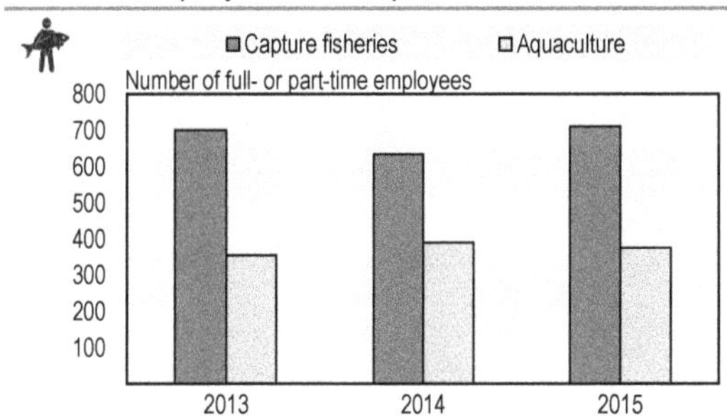

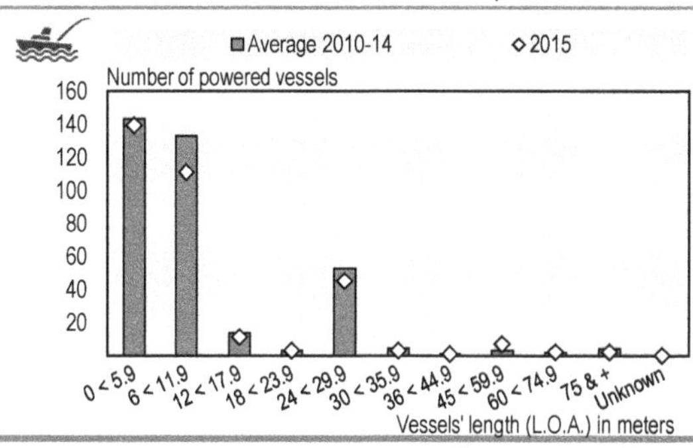

Government support and resource management

49% of total budgetary support was directed to management of resources in 2015

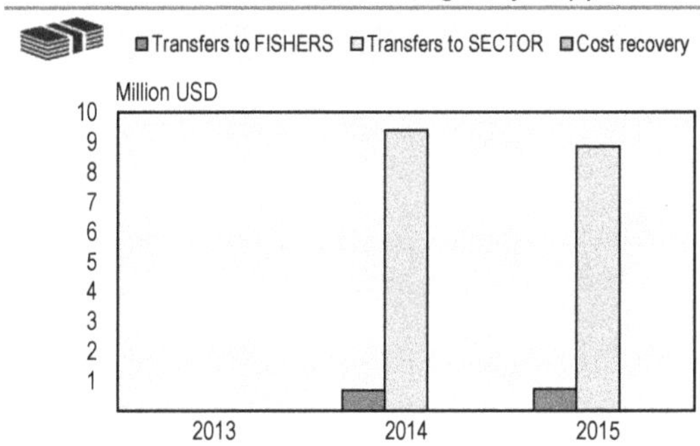

For information on EU fish stocks please refer to the EU country chapter

Note: Reference year is 2015. The 'Transfers to SECTOR' numbers reported here include estimations for management and enforcement expenditures, where missing.

Source: OECD Fisheries Database. World's total values of production for catch and aquaculture were sourced from the *FAO Yearbook, Fishery and Aquaculture Statistics, 2015*. Both numbers have been corrected adding the value of seaweed production; aquaculture has also been adjusted to account for the value of Chinese production reported to the OECD.

Lithuania

Developments in fisheries and aquaculture policy, 2015-16

- The new Fisheries Law extends the use of transferable rights to fishing quotas to the long-distance fishing fleet and small scale Baltic Sea fisheries. This extension is already in place for inland waters and the large-scale Baltic Sea fleet and now covers almost all fisheries and fleet segments.
- A new traceability system was created and came into force in early 2017. The system is based on electronic record-keeping and is implemented by the Fisheries Service under the Ministry of Agriculture.
- A target has been set to increase the production volume of aquaculture by 66% and the production value by 111% by 2022.

Fisheries and aquaculture statistics in Lithuania

Value of production

in global context

USD 83 million — *Total value of fisheries and aquaculture*

0.06% — *Share of global fisheries landed value*

0.006% — *Share of global aquaculture production value*

Trends in production value and trade

Total production value in USD increased at an annual average rate of 14.7% between 2011 and 2015

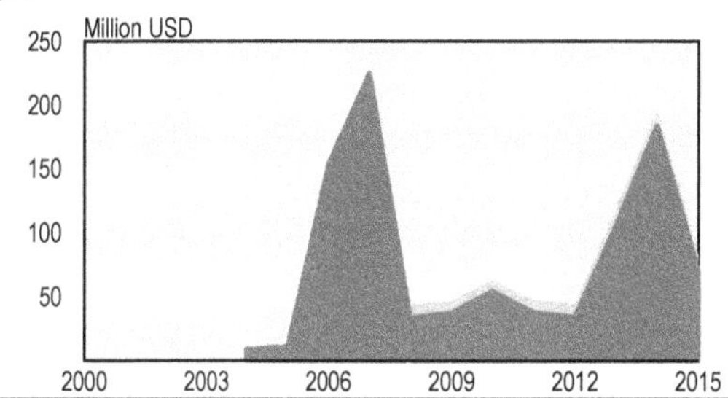

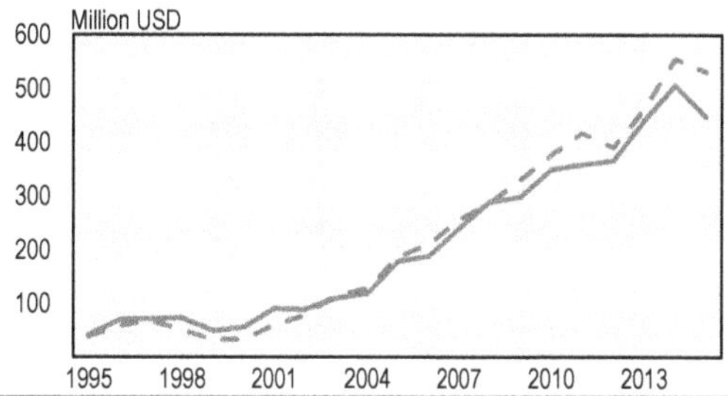

Employment and fleet

Employment in capture fisheries decreased by 7.9% between 2014 and 2015 to 691 persons

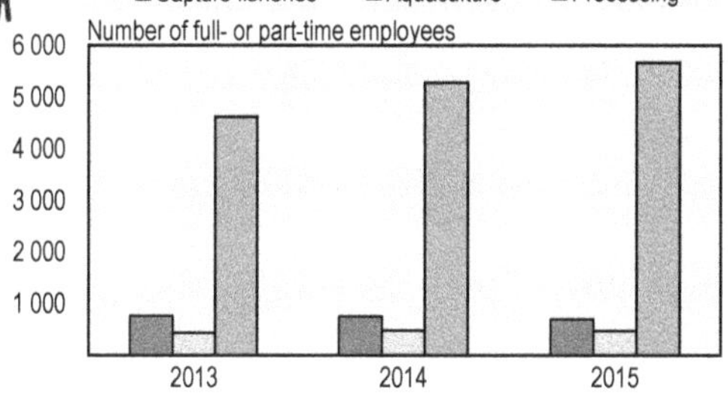

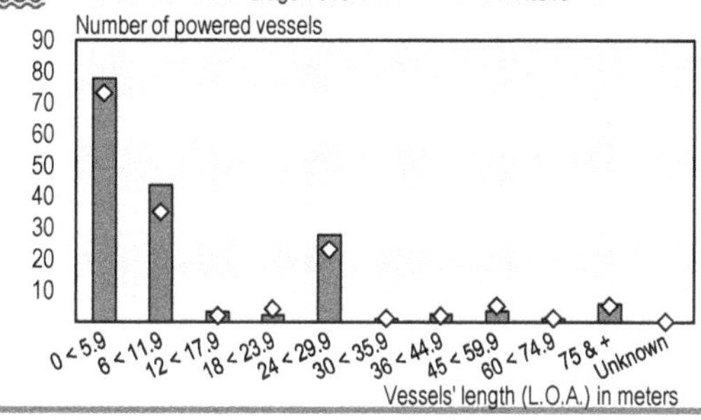

Government support and resource management

16% of total budgetary support was directed to management of resources in 2015

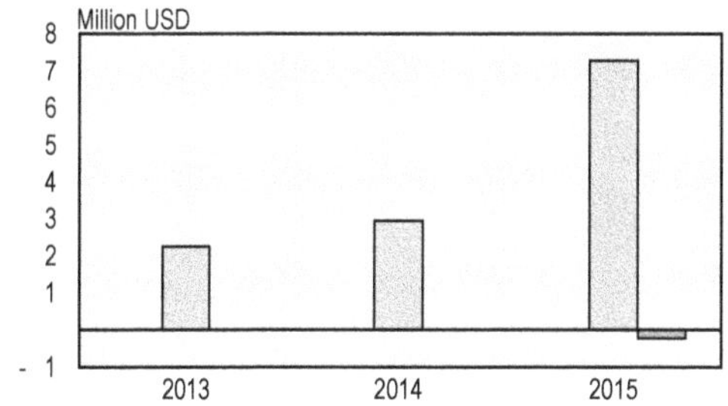

For information on EU fish stocks please refer to the EU country chapter

Note: Reference year is 2015.

Source: OECD Fisheries Database. World's total values of production for catch and aquaculture were sourced from the *FAO Yearbook, Fishery and Aquaculture Statistics, 2015*. Both numbers have been corrected adding the value of seaweed production; aquaculture has also been adjusted to account for the value of Chinese production reported to the OECD.

Mexico

Developments in fisheries and aquaculture policy, 2015-16

- The government of Mexico is currently implementing the National Development Plan 2013-2018 and the Sectorial Program of Farming and Fishing of the Secretariat of Agriculture, Livestock, Rural Development, Fisheries and Food. The programmes put an emphasis on the need for improved competitiveness in the fisheries and aquaculture sectors in the country.
- Mexico is in the process of restructuring its fisheries and aquaculture related regulations. The goal is to establish a procedure for regular updating of regulations in order to assure efficient conservation of available resources.
- Digitalisation plays an important role in the development of the aquaculture sector in Mexico. A small number of new web-based services are available. The goal is to support efficient operations, assure transparency in compliance with regulations and accelerate the permit application process.

Fisheries and aquaculture statistics in Mexico

Value of production

in global context

USD 1 984 million	1.1%	0.3%
Total value of fisheries and aquaculture	*Share of global fisheries landed value*	*Share of global aquaculture production value*

Trends in production value and trade

Total production value in USD increased at an annual average rate of 8.5% between 2011 and 2015

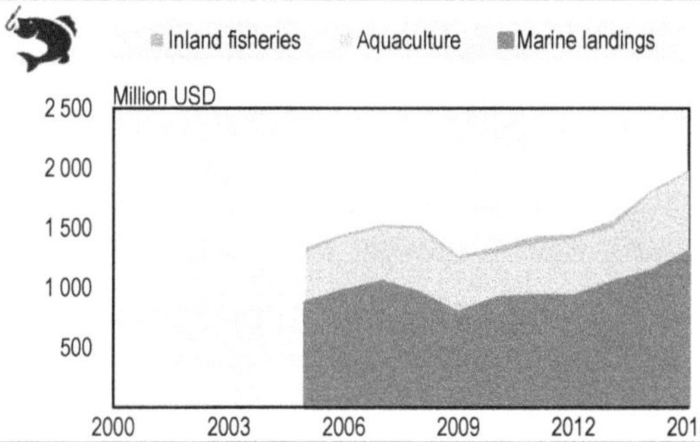

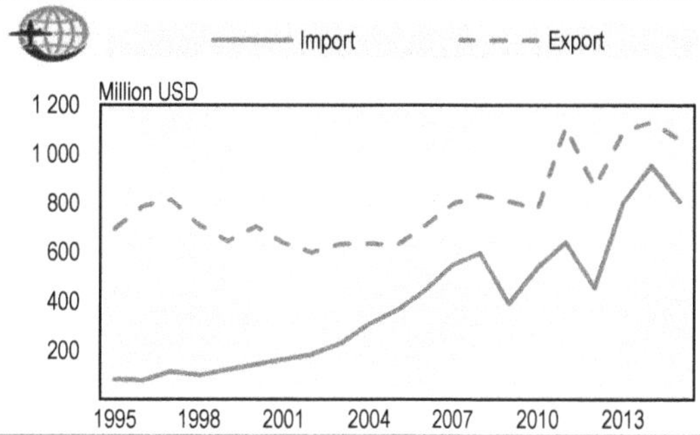

Employment and fleet

Employment in capture fisheries increased by 11.1% between 2014 and 2015 to 238 590 persons

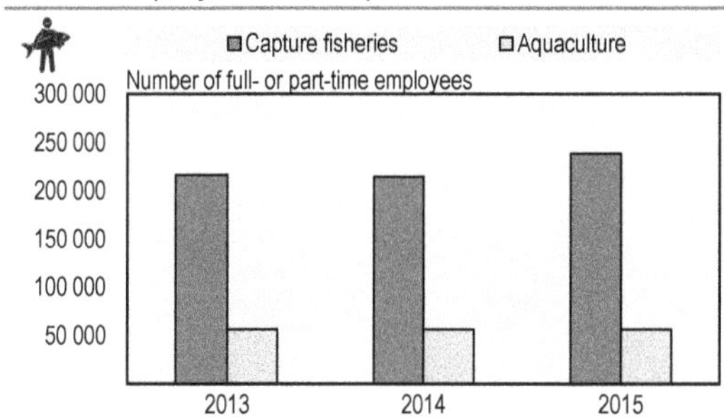

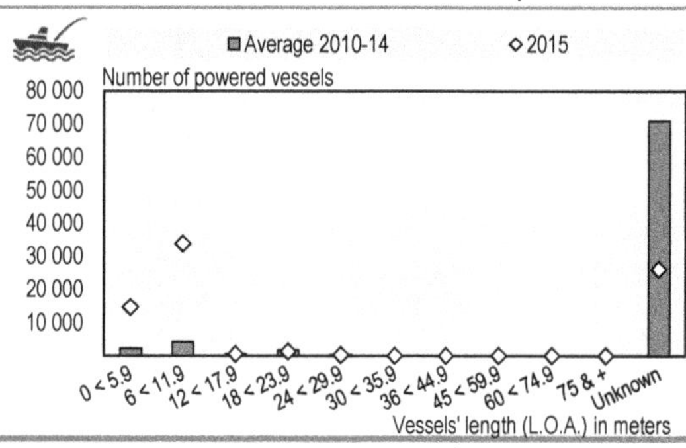

Government support and resource management

70% of total budgetary support was directed to management of resources in 2015

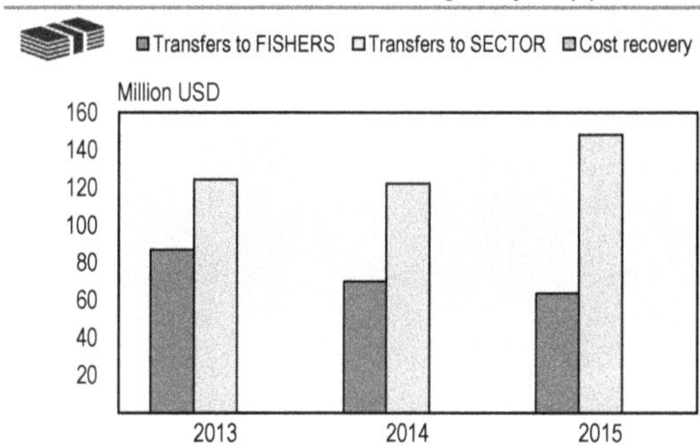

---	Number of stocks with established targets or thresholds recently subject to quantitative assessment
	of which
---	Number of stocks found to be within acceptable conditions

Note: Reference year is 2015. The 'Transfers to SECTOR' numbers reported here include estimations for management and enforcement expenditures, where missing.

Source: OECD Fisheries Database. World's total values of production for catch and aquaculture were sourced from the *FAO Yearbook, Fishery and Aquaculture Statistics, 2015*. Both numbers have been corrected adding the value of seaweed production; aquaculture has also been adjusted to account for the value of Chinese production reported to the OECD.

Netherlands

Developments in fisheries and aquaculture policy, 2015-16

- The economic performance of the Dutch fleet has improved and further profit growth is expected in 2015 and 2016 as a result of higher landings of valuable species, higher fish prices, lower energy costs due to decreasing fuel prices and use of fuel saving techniques.
- Financial support has been directed mainly towards data collection, better management and use of collected data, as well as the promotion of environmentally sustainable, resource-efficient, innovative, competitive and knowledge-based production techniques.
- Catch limits have been adopted for sea bass – for commercial fishing as well as for recreational fisheries – in the Atlantic and North Sea to prevent the stock from collapsing.

Fisheries and aquaculture statistics in the Netherlands

Value of production

in global context

USD 617 million	0.4%	0.06%
Total value of fisheries and aquaculture	*Share of global fisheries landed value*	*Share of global aquaculture production value*

Trends in production value and trade

Total production value in USD increased at an annual average rate of 1.8% between 2011 and 2014

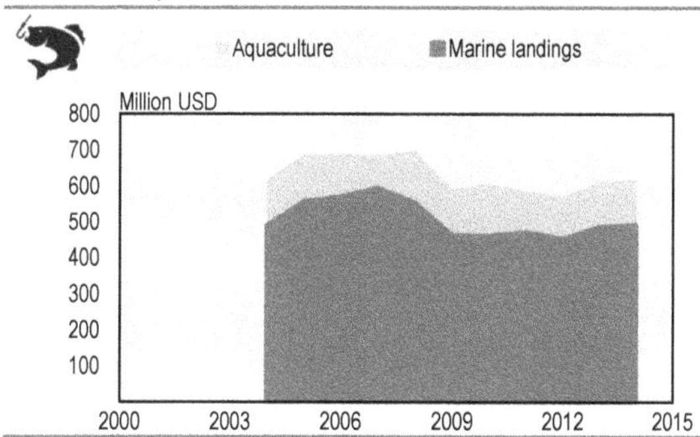

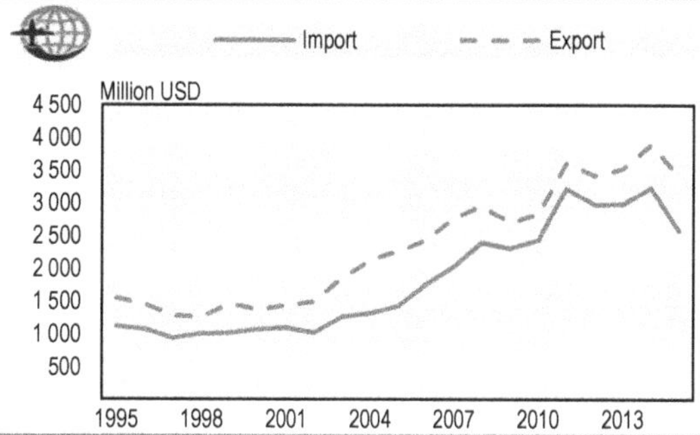

Employment and fleet

Employment in capture fisheries decreased by 8.4% between 2014 and 2015 to 1 694 persons

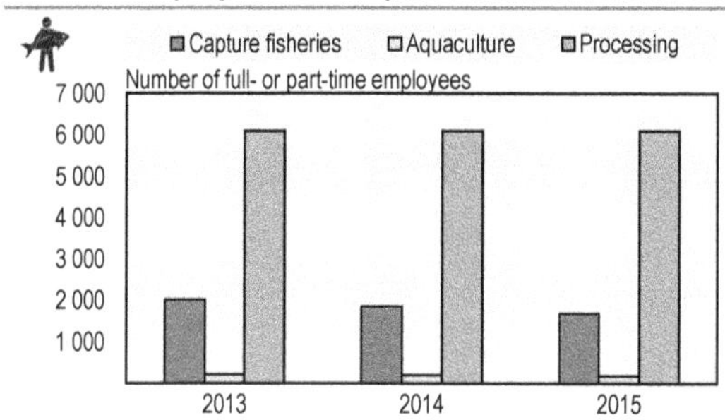

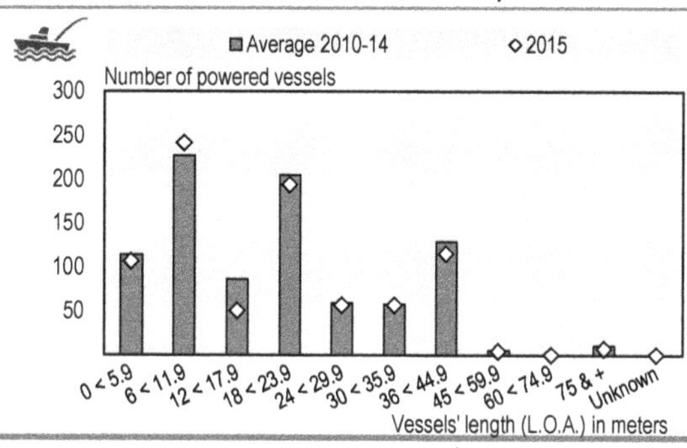

Government support and resource management

68% of total budgetary support was directed to management of resources in 2015

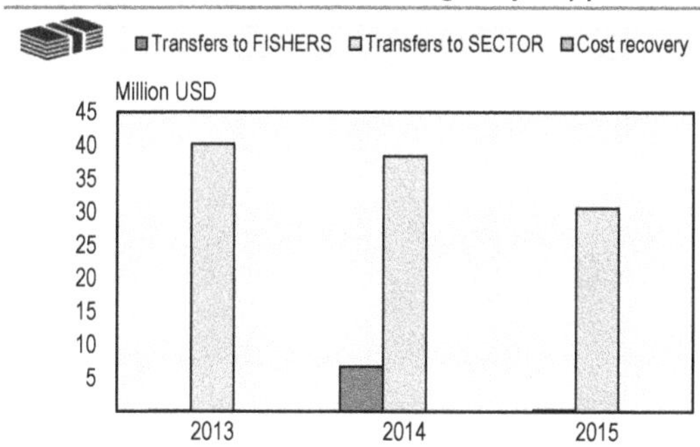

For information on EU fish stocks please refer to the EU country chapter

Note: Reference year is 2014. The 'Transfers to SECTOR' numbers reported here include estimations for management and enforcement expenditures, where missing.

Source: OECD Fisheries Database. World's total values of production for catch and aquaculture were sourced from the *FAO Yearbook, Fishery and Aquaculture Statistics, 2015*. Both numbers have been corrected adding the value of seaweed production; aquaculture has also been adjusted to account for the value of Chinese production reported to the OECD.

New Zealand

Developments in fisheries and aquaculture policy, 2015-16

- In 2015 the Ministry for Primary Industries (MPI) undertook a process of engagement to obtain public, stakeholder and Maori comment on the current operation of the fisheries management system. This had the objective of ensuring the management system is able to provide sustainable fisheries resources for all of New Zealand. Throughout 2016 a major work programme called Future of our Fisheries was developed on the basis of the feedback received and is now working towards informing proposed regulatory changes in the areas of electronic monitoring and trawl technology.
- Since 2016, all vessels operating within New Zealand's fisheries waters must be New Zealand flagged and are subject to all relevant domestic legislation.
- Marine Stewardship Council certification, which recognises the use of sustainable practices, has been extended to an additional three of New Zealand's fish stocks taking the total number to 21 stocks.
- The Marine Protected Areas Act was proposed during this period, which, if adopted, would enhance, protect and restore marine biodiversity in its coastal marine area by taking a more integrated policy approach to the management of marine resources.

Fisheries and aquaculture statistics in New Zealand

Value of production

in global context

USD 1 183 million	0.4%	0.3%
Total value of fisheries and aquaculture	*Share of global fisheries landed value*	*Share of global aquaculture production value*

Trends in production value and trade

Total production value in USD increased at an annual average rate of 8.5% between 2011 and 2015

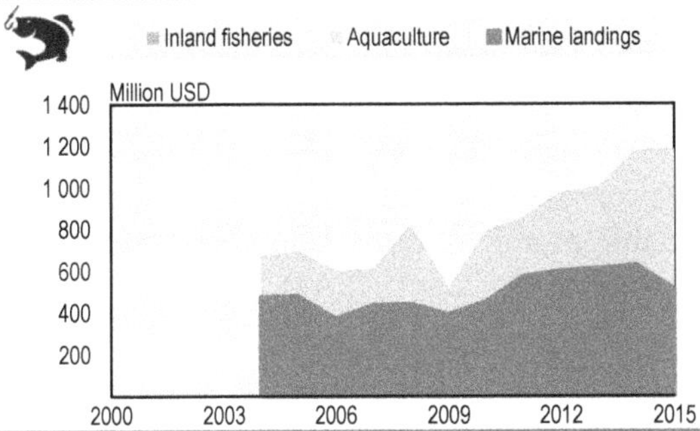

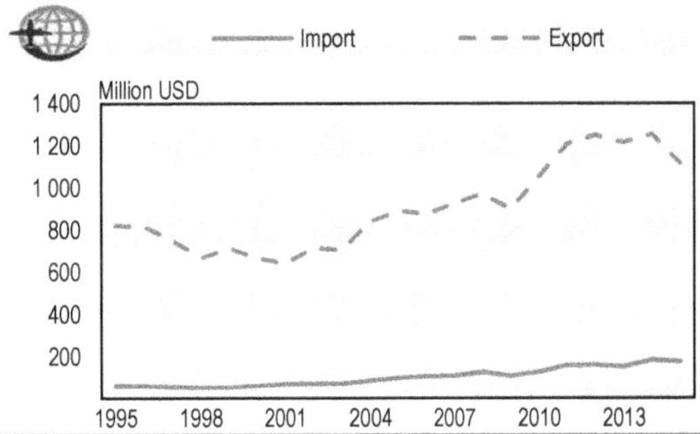

Employment and fleet

Employment in capture fisheries increased by 10.8% between 2014 and 2015 to 2 050 persons

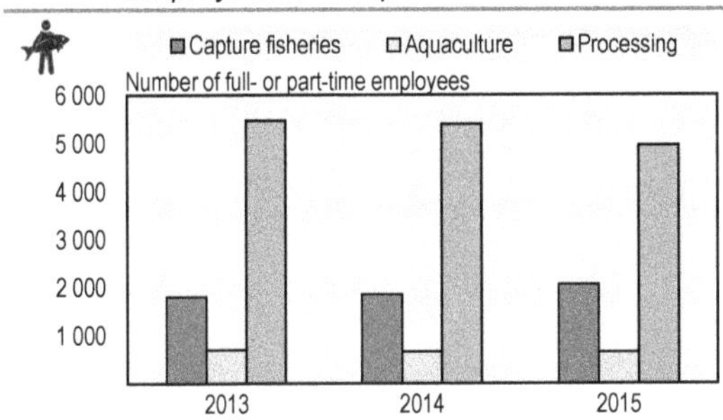

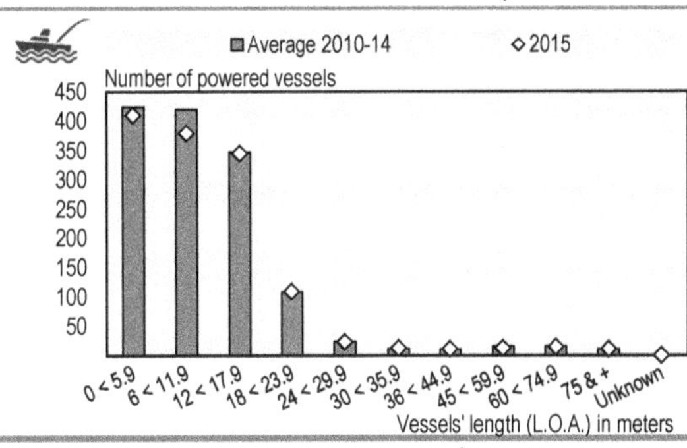

Government support and resource management

75% of total budgetary support was directed to management of resources in 2015

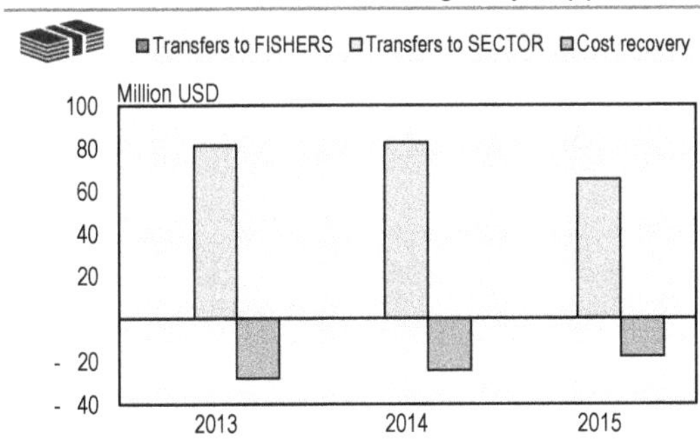

161 Number of stocks with established targets or thresholds recently subject to quantitative assessment

of which

134 Number found to have acceptable status with respect to targets or thresholds

Note: Reference year is 2015, except for stock assessment data which may cover more than one year. The 'Transfers to SECTOR' numbers reported here include an estimate for management expenditure in 2013.
Source: OECD Fisheries Database. World's total values of production for catch and aquaculture were sourced from the *FAO Yearbook, Fishery and Aquaculture Statistics, 2015*. Both numbers have been corrected adding the value of seaweed production; aquaculture has also been adjusted to account for the value of Chinese production reported to the OECD. Fish stocks data were directly reported by the country and are available from The status of New Zealand Fisheries 2015.

Norway

Developments in fisheries and aquaculture policy, 2015-16

- A new system for regulating growth in the salmon farming industry was introduced in 2015 and implemented in 2017 based on environmental indicators and production areas. Furthermore, Innovation Licences were introduced, a hybrid of research licences and regular commercial licences aimed at stimulating productivity-enhancing innovations.
- To reduce complexity, a plan to reform the existing quota management system for capture fisheries is under consideration. A resource rent taxation scheme will be part of this process.
- Area-based fisheries management is being progressively introduced, including during this reporting period, to ensure sustainable production and reduce resource conflicts, including through the use of Marine Protected Areas (MPAs).

Fisheries and aquaculture statistics in Norway

Value of production

in global context

USD 7 918 million	1.7%	2.7%
Total value of fisheries and aquaculture	*Share of global fisheries landed value*	*Share of global aquaculture production value*

Trends in production value and trade

Total production value in USD decreased at an annual average rate of 0.4% between 2011 and 2015

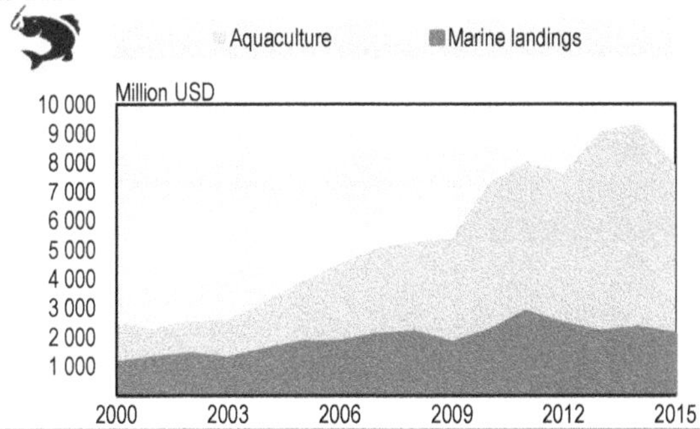

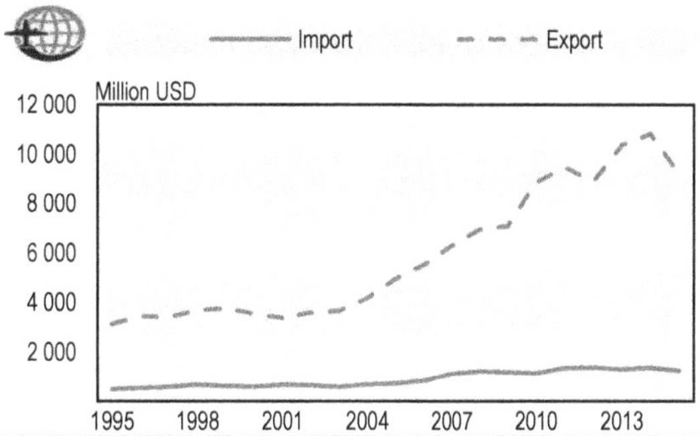

Employment and fleet

Employment in capture fisheries decreased by 1.5% between 2014 and 2015 to 11 130 persons

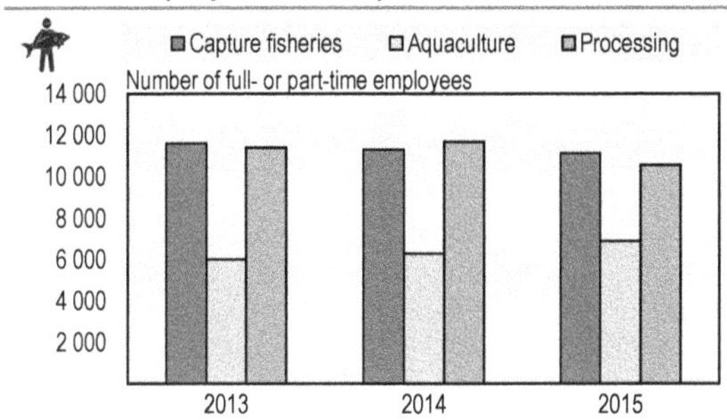

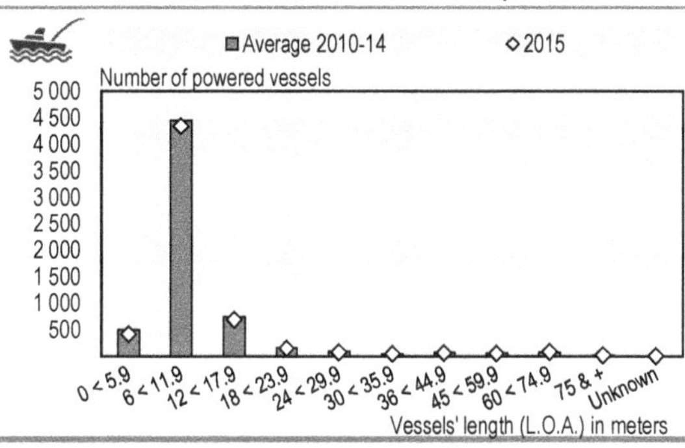

Government support and resource management

55% of total budgetary support was directed to management of resources in 2015

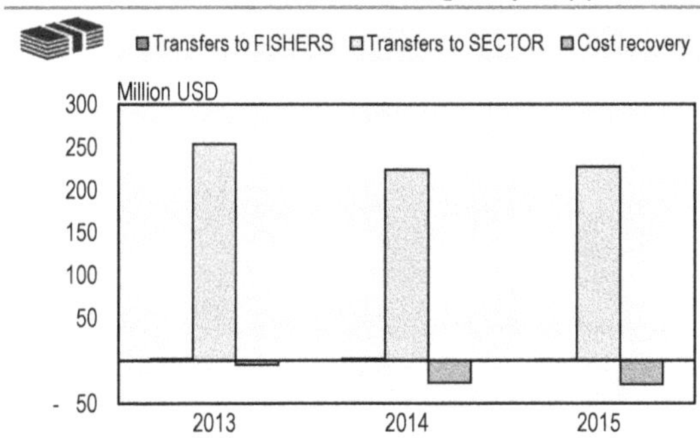

75 Number of stocks with established targets or thresholds recently subject to quantitative assessment

of which

47 Number found to have acceptable status with respect to targets or thresholds

Note: Reference year is 2015, except for stock assessment data which may cover more than one year.
Source: OECD Fisheries Database. World's total values of production for catch and aquaculture were sourced from the *FAO Yearbook, Fishery and Aquaculture Statistics, 2015*. Both numbers have been corrected adding the value of seaweed production; aquaculture has also been adjusted to account for the value of Chinese production reported to the OECD. Fish stocks data were directly reported by the country.

Poland

Developments in fisheries and aquaculture policy, 2015-16

- Polish catches from the Baltic Sea have consistently increased during the review period, but the composition of these catches is changing. Cod has traditionally been the most valuable catch but, has recently been replaced by European flounder due to poor cod stock status. Pelagic species, herring and sprat constituted the most valuable fisheries in 2016.
- The largest share of support from the European Maritime and Fisheries Fund allocated to Poland is intended for the development of aquaculture. Poland aims to become a European Union leader in intensive aquaculture production.
- The assessment of the Baltic Sea environment, where Polish fisheries are concentrated, conducted under the HELCOM HOLAS II project produced mixed results. Hazardous substance pollution is decreasing, but eutrophication remains a key concern. Only three out of eight assessed stocks were classified as being in good status.

Fisheries and aquaculture statistics in Poland

Value of production

in global context

USD 159 million	0.1%	0.05%
Total value of fisheries and aquaculture	*Share of global fisheries landed value*	*Share of global aquaculture production value*

Trends in production value and trade

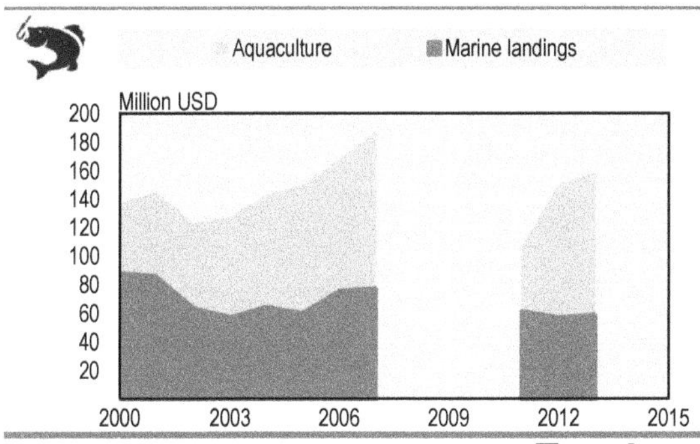

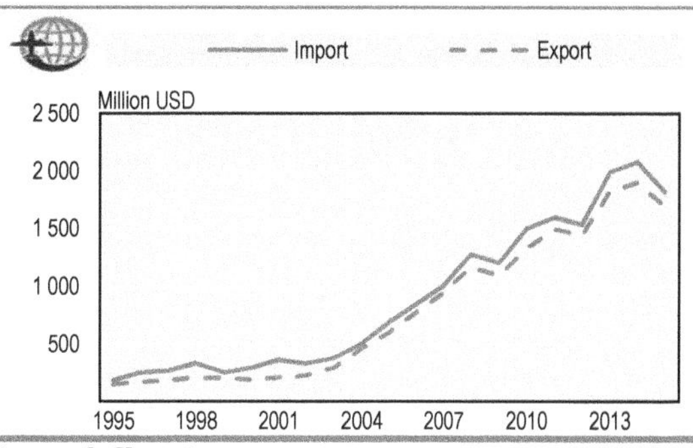

Employment and fleet

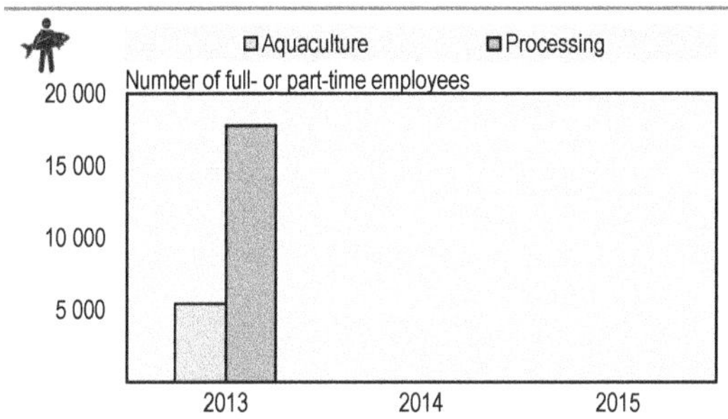

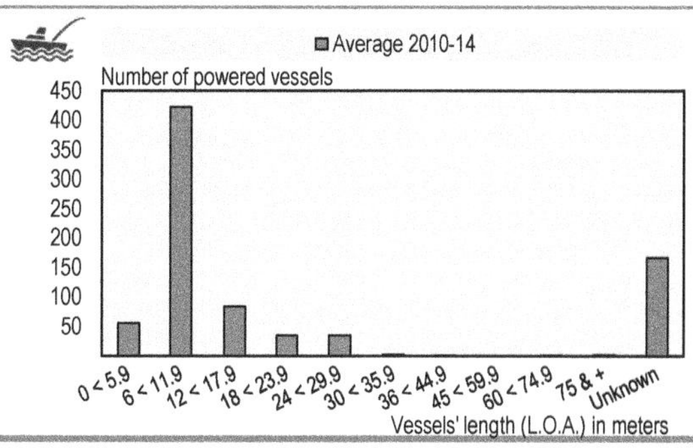

Government support and resource management

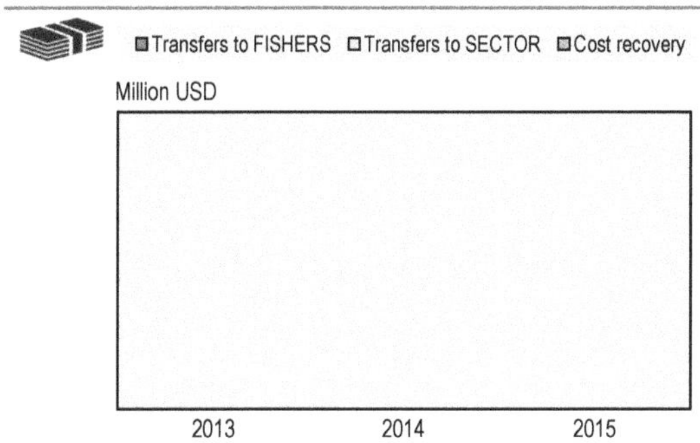

For information on EU fish stocks please refer to the EU country chapter

Note: Reference year is 2013.

Source: OECD Fisheries Database. World's total values of production for catch and aquaculture were sourced from the *FAO Yearbook, Fishery and Aquaculture Statistics, 2015*. Both numbers have been corrected adding the value of seaweed production; aquaculture has also been adjusted to account for the value of Chinese production reported to the OECD.

Portugal

Developments in fisheries and aquaculture policy, 2015-16

- Aquaculture output is increasing in Portugal and with support from the European Maritime and Fisheries Fund it aims to treble the volume produced while also increasing the diversity of species farmed by 2023.
- Sustainable development objectives have resulted in deep sea habitats being protected from trawling, and laws enacted in 2014 pave the way for sustainable spatial planning and aquaculture development.
- The Portuguese fishing sector has recognised the need for management measures and is taking a greater role in the decision-making process.

Fisheries and aquaculture statistics in Portugal

Value of production

in global context

USD 470 million	0.3%	0.03%
Total value of fisheries and aquaculture	*Share of global fisheries landed value*	*Share of global aquaculture production value*

Trends in production value and trade

Total production value in USD decreased at an annual average rate of 8.5% between 2011 and 2015

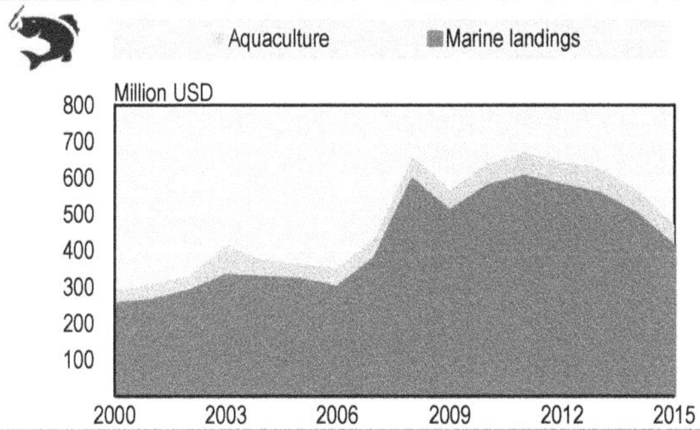

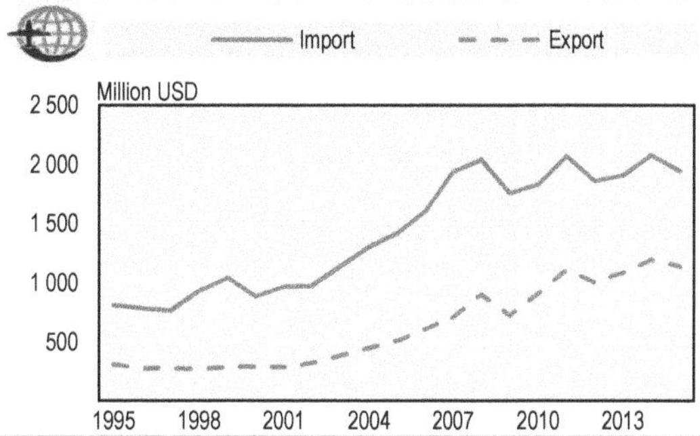

Employment and fleet

Employment in capture fisheries increased by 4.5% between 2014 and 2015 to 17 536 persons

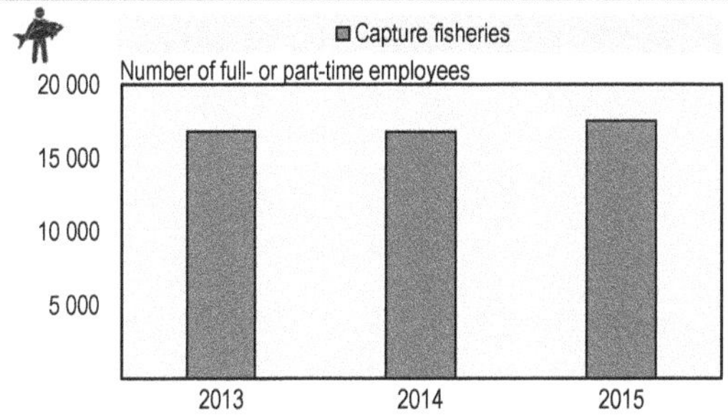

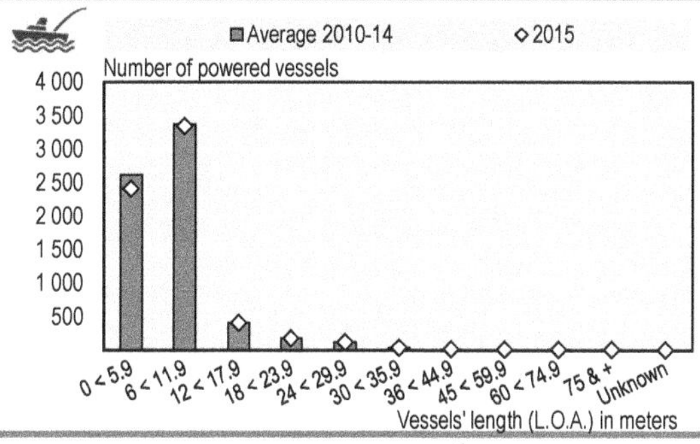

Government support and resource management

94% of total budgetary support was directed to management of resources in 2015

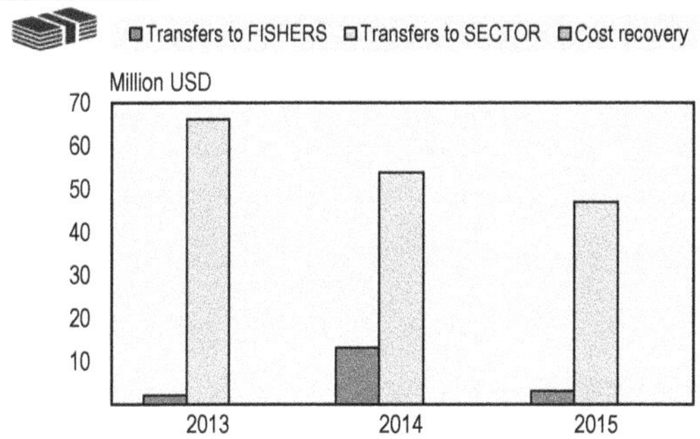

For information on EU fish stocks please refer to the EU country chapter

Note: Reference year is 2015. The 'Transfers to SECTOR' numbers reported here include estimations for management and enforcement expenditures, where missing.

Source: OECD Fisheries Database. World's total values of production for catch and aquaculture were sourced from the *FAO Yearbook, Fishery and Aquaculture Statistics, 2015*. Both numbers have been corrected adding the value of seaweed production; aquaculture has also been adjusted to account for the value of Chinese production reported to the OECD.

Slovak Republic

Developments in fisheries and aquaculture policy, 2015-16

- The Slovak Republic plans to simplify its current certification scheme of aquaculture farms. The goal is to ease access to the sector and encourage new entrants.
- The main objective of investments in the aquaculture sector is to increase domestic fish supply to 80% in volume by 2020, up from current level of 40%. At the same time, funds from the European Maritime and Fisheries Fund support a move towards sustainable production which takes into account environmental, social and economic aspects.

Fisheries and aquaculture statistics in the Slovak Republic

Value of production

in global context

USD 4 million	---	0.002%
Total value of fisheries and aquaculture	*Share of global fisheries landed value*	*Share of global aquaculture production value*

Trends in production value and trade

Total production value in USD increased at an annual average rate of 3.6% between 2011 and 2015

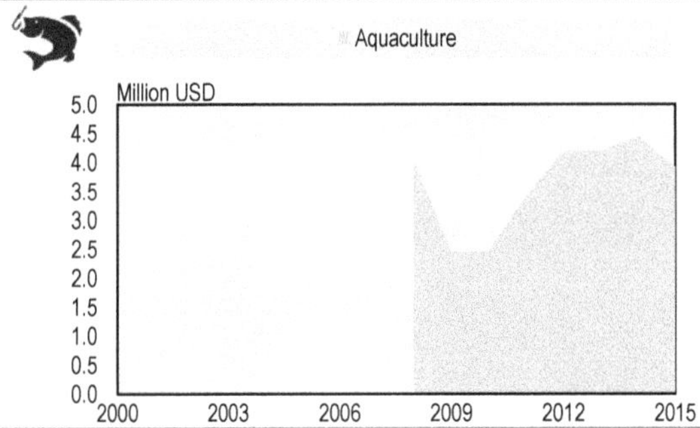

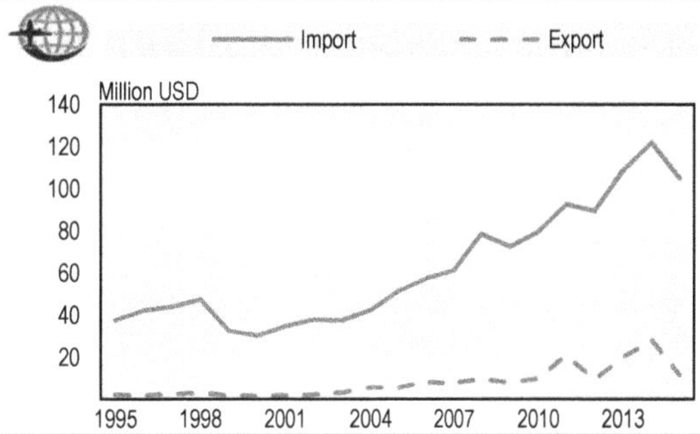

Employment and fleet

Employment in aquaculture increased by 73.5% between 2014 and 2015 to 734 persons

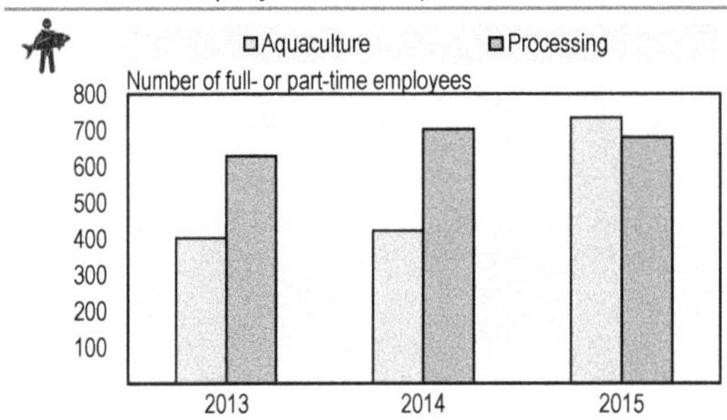

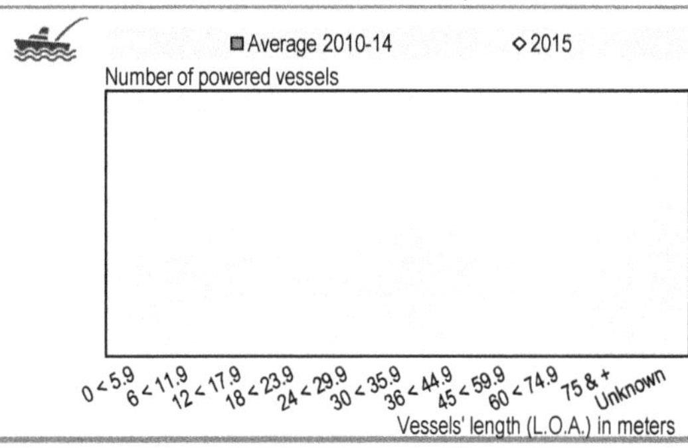

Government support and resource management

There is no marine fisheries production in the country

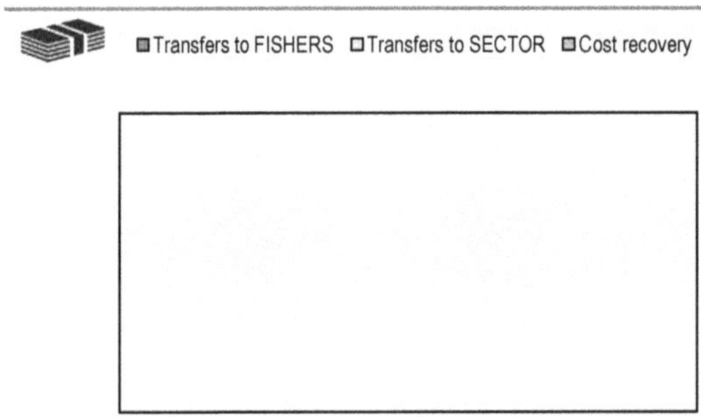

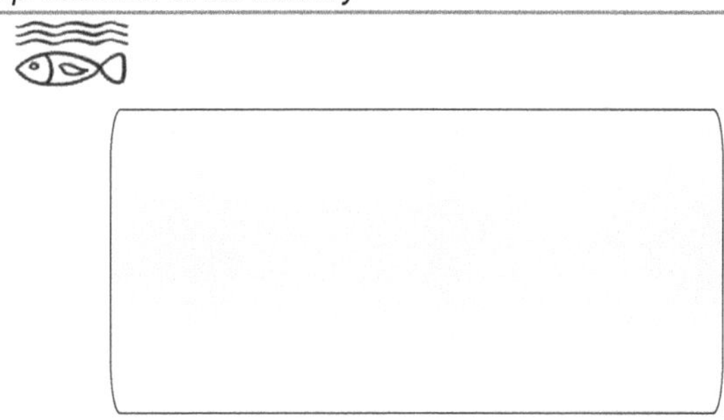

Note: Reference year is 2015.

Source: OECD Fisheries Database. World's total values of production for catch and aquaculture were sourced from the *FAO Yearbook, Fishery and Aquaculture Statistics, 2015*. Both numbers have been corrected adding the value of seaweed production; aquaculture has also been adjusted to account for the value of Chinese production reported to the OECD.

Slovenia

Developments in fisheries and aquaculture policy, 2015-16

- Slovenia's Operational Programme for 2014-20 allocates the largest share of funds for aquaculture development.
- The poor state of anchovy and sardine stocks in the Adriatic Sea has led to actions to reduce capacity and effort and to preparation of new management plans related to the status of these stocks.
- The government is concerned that reduced fishing opportunities, due to declining stocks, in the Adriatic Sea pose a risk to small-scale fishers.

Fisheries and aquaculture statistics in Slovenia

Value of production

in global context

USD 5 million	0.001%	0.002%
Total value of fisheries and aquaculture	*Share of global fisheries landed value*	*Share of global aquaculture production value*

Trends in production value and trade

Fisheries landed value in USD decreased at an average annual rate of 10% between 2011 and 2015

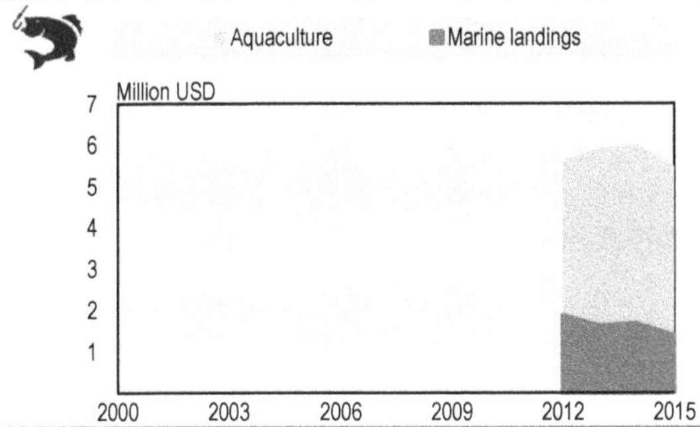

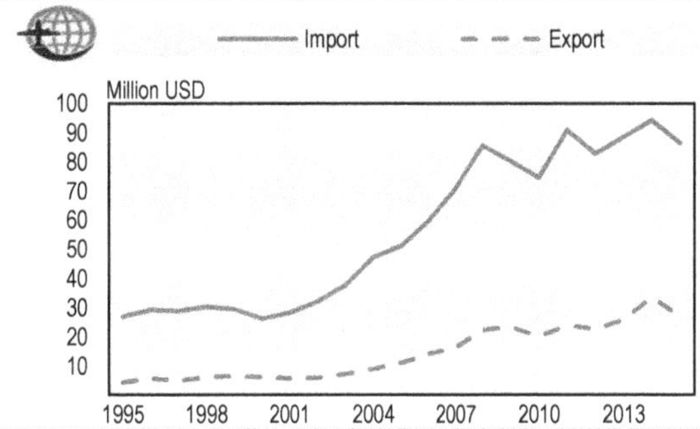

Employment and fleet

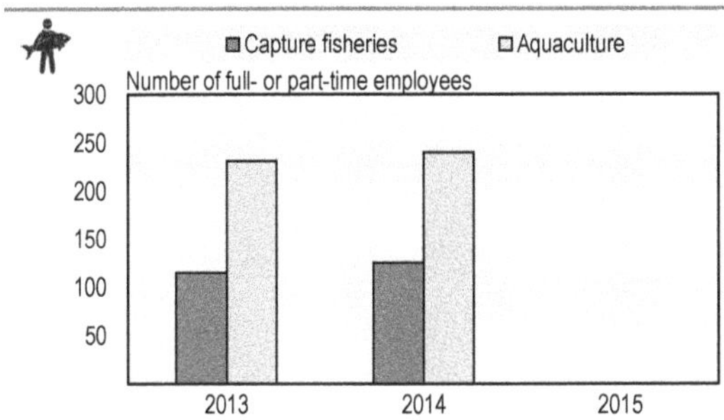

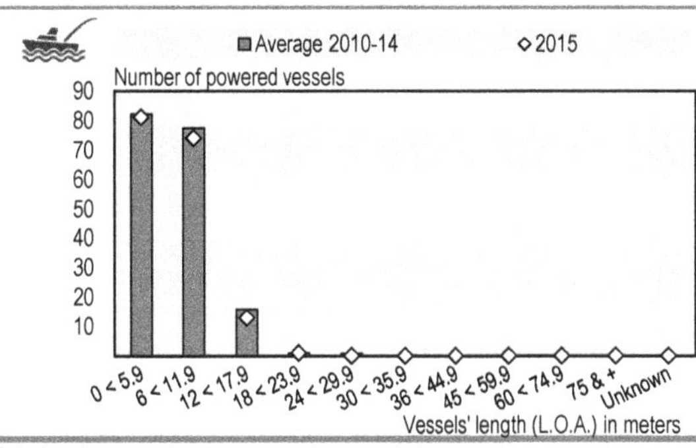

Government support and resource management

16% of total budgetary support was directed to management of resources in 2015

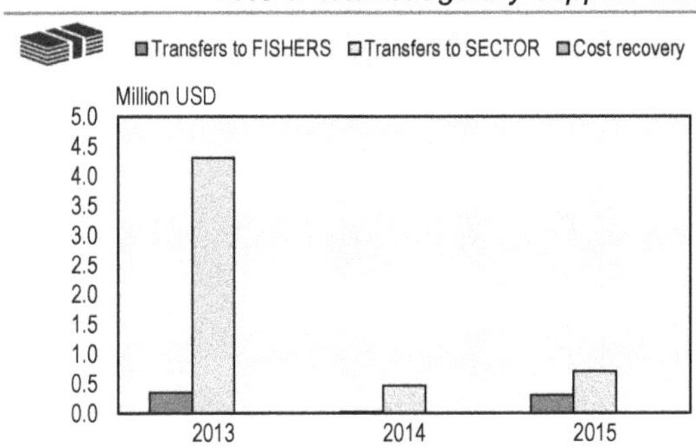

For information on EU fish stocks please refer to the EU country chapter

Note: Reference year is 2015. The 'Transfers to SECTOR' numbers reported here include estimations for management and enforcement expenditures, where missing.

Source: OECD Fisheries Database. World's total values of production for catch and aquaculture were sourced from the *FAO Yearbook, Fishery and Aquaculture Statistics, 2015*. Both numbers have been corrected adding the value of seaweed production; aquaculture has also been adjusted to account for the value of Chinese production reported to the OECD.

Spain

Developments in fisheries and aquaculture policy, 2015-16

- Spain's investments in fisheries are focused on adapting the fleet to evolving environmental conditions and protecting aquatic biodiversity. Concern about marine ecosystems has resulted in the creation of new marine protected areas (MPAs) and policies to address the impact of fisheries on the marine environment.
- The national strategic plan for aquaculture aims to boost the competitiveness of Spain's fish farming sector. To this end, detailed plans have been developed to maintain environmental integrity along the production chain.
- Spain is also working towards more transparent aquaculture production by implementing a unified information system that traces seafood origin. In addition, efforts to provide industry with a well-functioning certification scheme are underway.
- Spain is also investing in research to better understand the potential role of the aquaculture industry in global climate change.

Fisheries and aquaculture statistics in Spain

Value of production

in global context

USD 4 277 million	2.9%	0.3%
Total value of fisheries and aquaculture	*Share of global fisheries landed value*	*Share of global aquaculture production value*

Trends in production value and trade

Total production value in USD increased at an annual average rate of 8.4% between 2011 and 2014

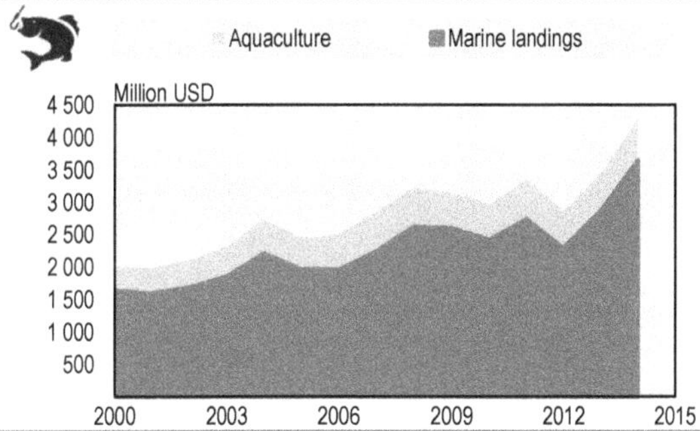

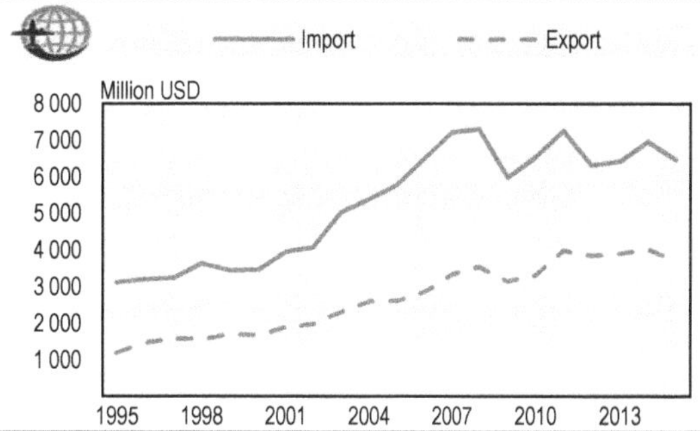

Employment and fleet

Employment in capture fisheries decreased by 3.4% between 2014 and 2015 to 25 863 persons

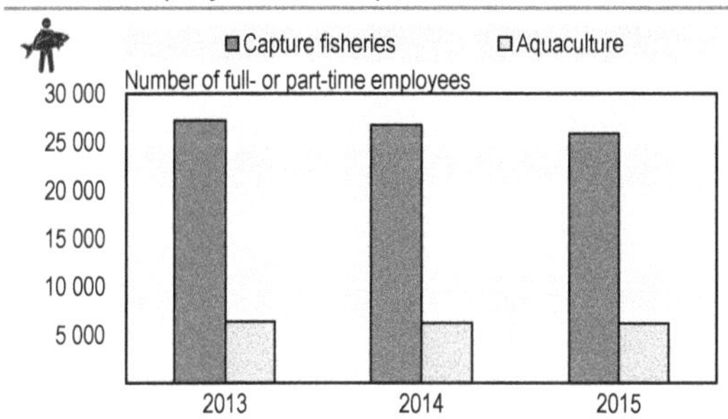

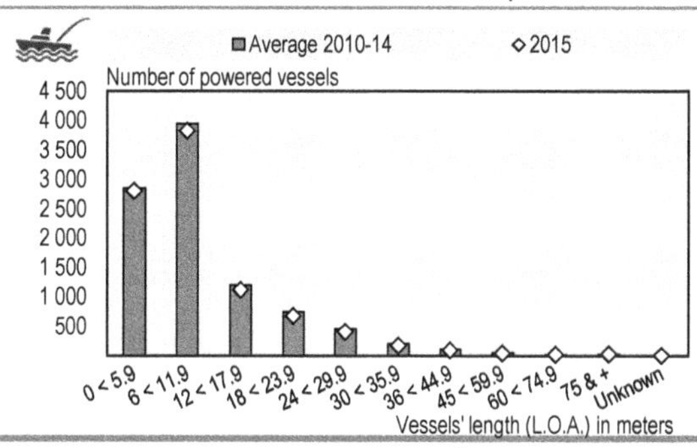

Government support and resource management

67% of total budgetary support was directed to management of resources in 2015

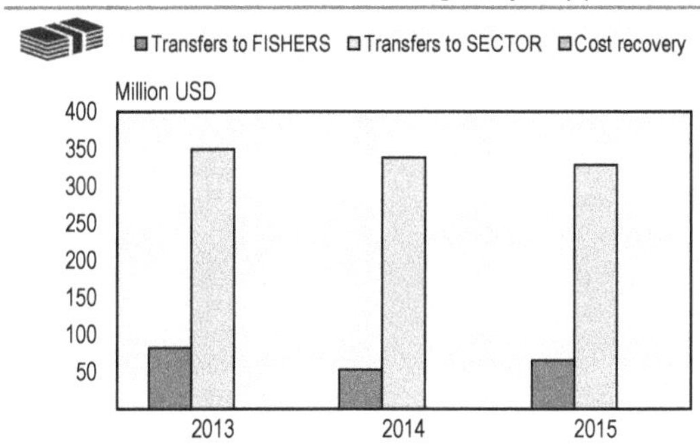

For information on EU fish stocks please refer to the EU country chapter

Note: Reference year is 2014. The 'Transfers to SECTOR' numbers reported here include estimations for management and enforcement expenditures, where missing.

Source: OECD Fisheries Database. World's total values of production for catch and aquaculture were sourced from the *FAO Yearbook, Fishery and Aquaculture Statistics, 2015*. Both numbers have been corrected adding the value of seaweed production; aquaculture has also been adjusted to account for the value of Chinese production reported to the OECD.

Sweden

Developments in fisheries and aquaculture policy, 2015-16

- Sweden tackles landing obligations by introducing limited transferability of fishing rights in the demersal sector. More flexibility in the system aims at better utilisation of national quotas.
- Land-based aquaculture production increased in Sweden. The changes in the sector are demand driven, following a growing interest in sustainable production techniques.
- The Swedish Government actively supports work on a strategy to operationalise an ecosystem-based fisheries management. The effort aligns with the goal of Good Environmental Status by 2020, following the definition from the Marine Strategy Framework Directive.

Fisheries and aquaculture statistics in Sweden

Value of production

in global context

USD 186 million	0.1%	0.02%
Total value of fisheries and aquaculture	*Share of global fisheries landed value*	*Share of global aquaculture production value*

Trends in production value and trade

Total production value in USD decreased at an annual average rate of 6.6% between 2011 and 2015

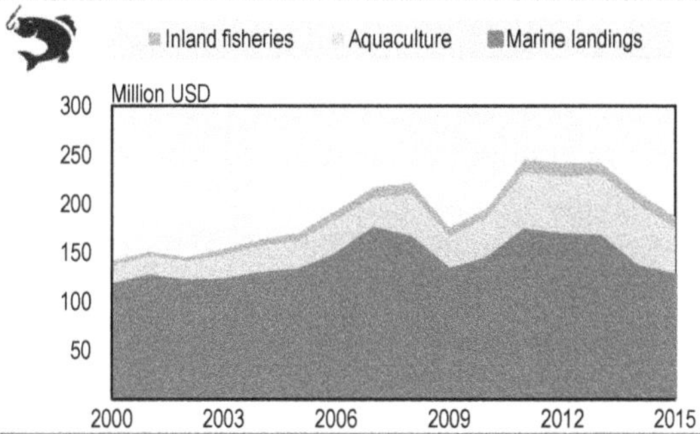

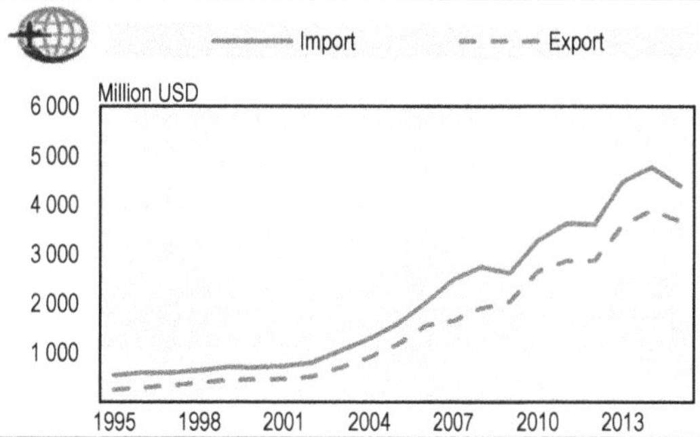

Employment and fleet

Employment in capture fisheries decreased by 0.7% between 2014 and 2015 to 1 657 persons

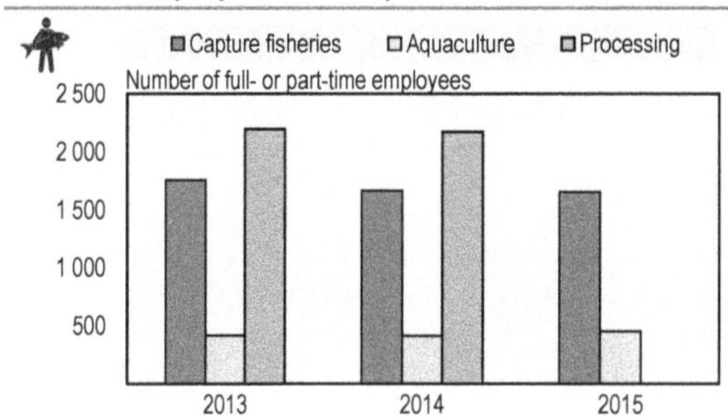

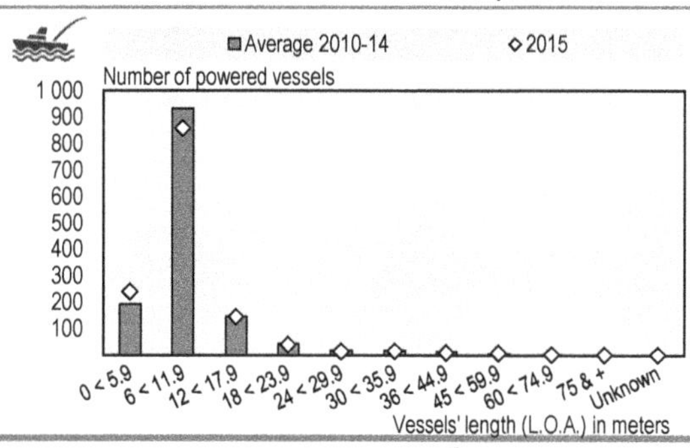

Government support and resource management

57% of total budgetary support was directed to management of resources in 2015

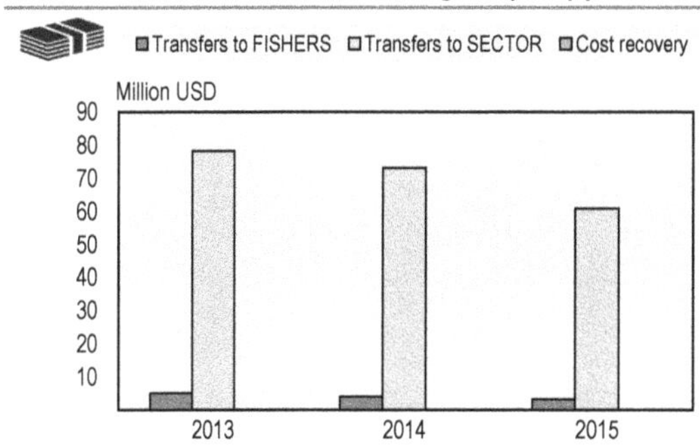

For information on EU fish stocks please refer to the EU country chapter

Note: Reference year is 2015.

Source: OECD Fisheries Database. World's total values of production for catch and aquaculture were sourced from the *FAO Yearbook, Fishery and Aquaculture Statistics, 2015*. Both numbers have been corrected adding the value of seaweed production; aquaculture has also been adjusted to account for the value of Chinese production reported to the OECD.

Chinese Taipei

Developments in fisheries and aquaculture policy, 2015-16

- The Five-Year Program for strengthening international co-operation on combating IUU fishing began in January 2016. It contains several elements, including a legal framework, an international economic and trade strategy, rules for traceability of fish and fisheries products, new regulations for control and management of coastal and offshore fishing vessels in port and at sea.
- Policies and fisheries law have been reinforced in 2015 and 2016 to address overfishing, overcapacity and IUU fishing. This includes better management of vessels operating in different fisheries as well as regulations adopted by RFMOs to implement effective measures to combat IUU fishing.
- Major policy changes include the establishment of a National Plan of Action - Illegal, Unreported, and Unregulated Fishing (NPOA-IUU), reinforcing monitoring, control, and surveillance (MCS) measures, enhancing catch traceability and promoting of sustainable aquaculture and traceability of fisheries products.

Fisheries and aquaculture statistics in Chinese Taipei

Value of production

in global context

USD 2 867 million	1.4%	0.6%
Total value of fisheries and aquaculture	*Share of global fisheries landed value*	*Share of global aquaculture production value*

Trends in production value and trade

Total production value in USD decreased at an annual average rate of 5.3% between 2011 and 2015

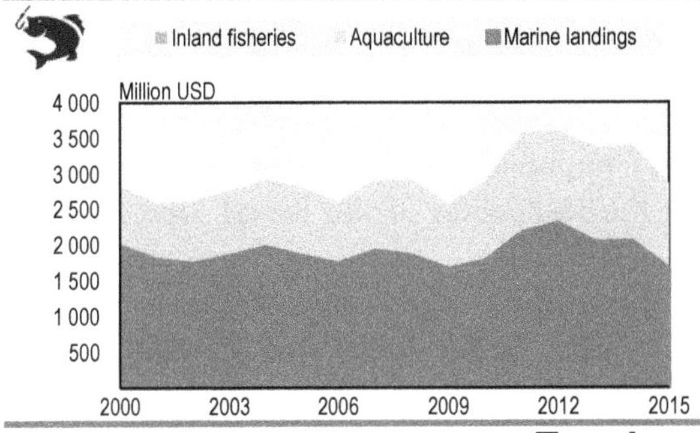

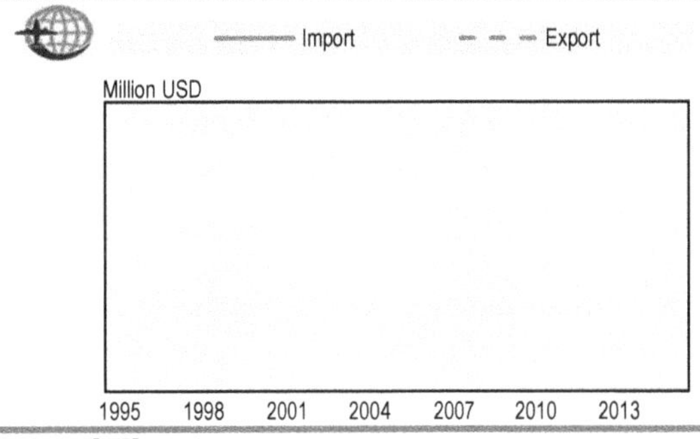

Employment and fleet

Employment in capture fisheries decreased by 3.1% between 2014 and 2015 to 236 593 persons

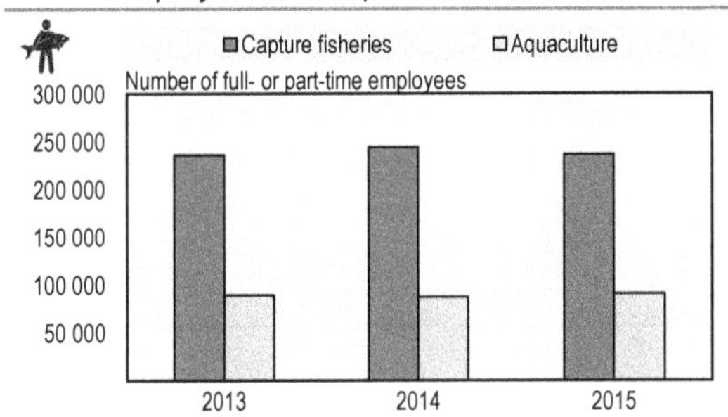

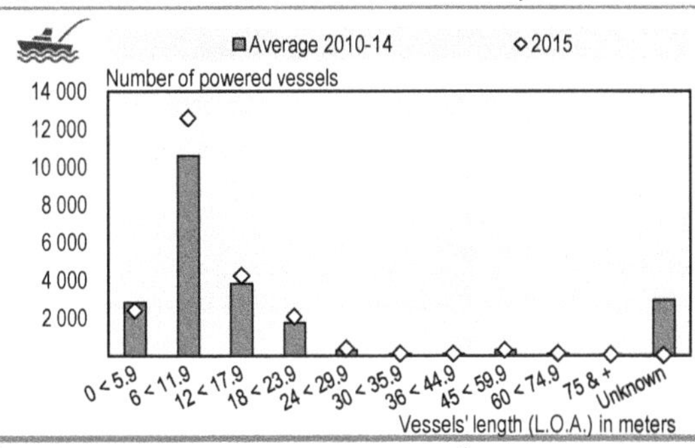

Government support and resource management

48% of total budgetary support was directed to management of resources in 2015

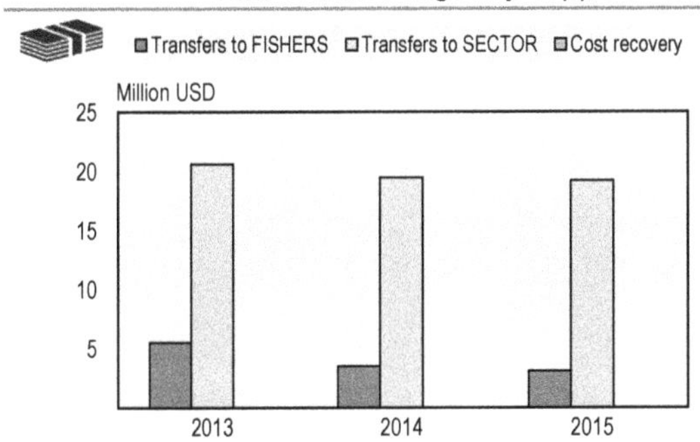

25 Number of stocks with established targets or thresholds recently subject to quantitative assessment

of which

15 Number found to have acceptable status with respect to targets or thresholds

Note: Reference year is 2015, except for stock assessment data which may cover more than one year.
Source: OECD Fisheries Database. World's total values of production for catch and aquaculture were sourced from the *FAO Yearbook, Fishery and Aquaculture Statistics, 2015*. Both numbers have been corrected adding the value of seaweed production; aquaculture has also been adjusted to account for the value of Chinese production reported to the OECD. Fish stocks data were directly reported by the country.

Thailand

Developments in fisheries and aquaculture policy, 2015-16

In 2015 and 2016, Thailand's governance and management framework for fisheries and aquaculture was structurally reformed to promote sustainable and responsible practices throughout the sector. Key changes include:

- Adoption of the Royal Ordinance on Fisheries (ROF) B.E. 2558, which entered into force in November 2015 and aims to empower relevant authorities to combat illegal, unreported and unregulated (IUU) fishing and unlawful labour practices in the fishing and seafood industries. A National Plan of Action to Prevent, Deter and Eliminate IUU Fishing was also adopted while a Command Centre for Combating Illegal Fishing (CCCIF) was established under the leadership of the Royal Thai Navy.
- Adoption of a Marine Fisheries Management Plan (MFP), which aims to tackle overfishing and overcapacity of the Thai fishing fleet. In the framework of this plan, a ban on new vessel registration was established in August 2015, as well as a vessel buy-back scheme, more stringent gear regulations, limits on days at sea and Total Allowable Catch (TAC) limits with a maximum sustainable yield (MSY) objective.
- Adoption of a National Plan of Control and Inspection (NPCI), which sets new requirements in terms of the use of International Maritime Organisation (IMO) numbers, Vessel Monitoring Systems (VMS) and a temporary ban on transhipment.

Fisheries and aquaculture statistics in Thailand

Value of production

in global context

USD 4 483 million	1.5%	1.2%
Total value of fisheries and aquaculture	*Share of global fisheries landed value*	*Share of global aquaculture production value*

Trends in production value and trade

Between 2012 (peak year) and 2014 production and trade declined by respectively 20% and 10%

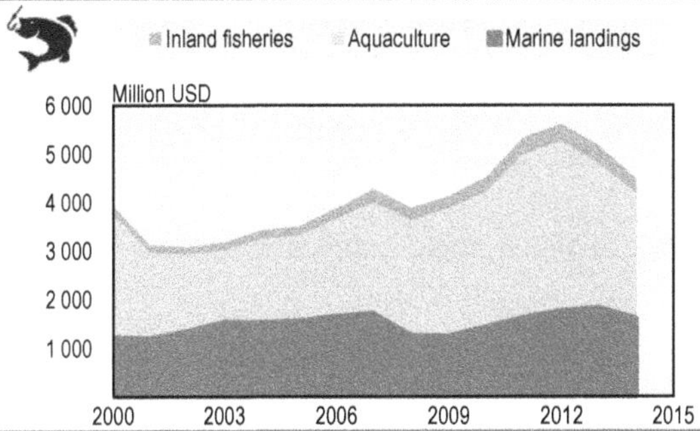

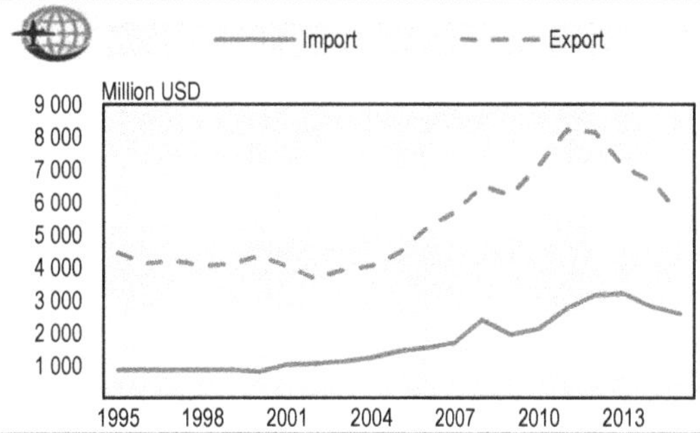

Employment and fleet

Employment in capture fisheries increased by 64.6% between 2014 and 2015 to 99 246 persons

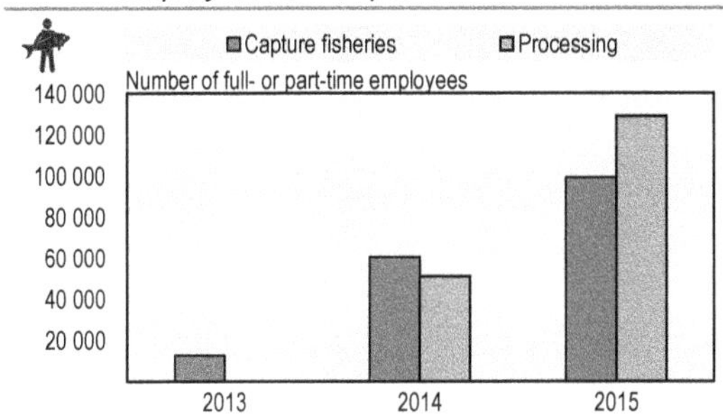

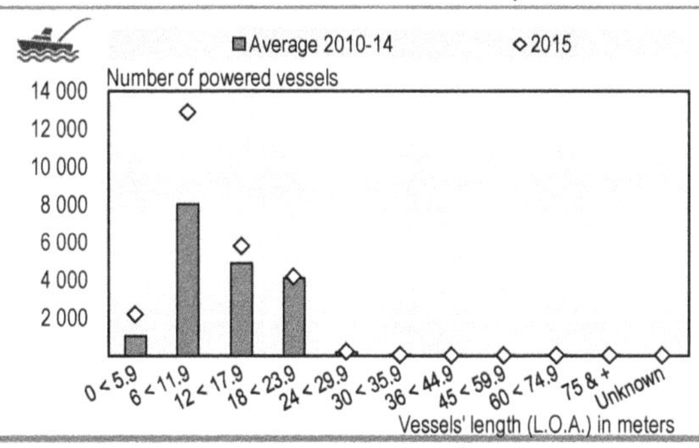

Government support and resource management

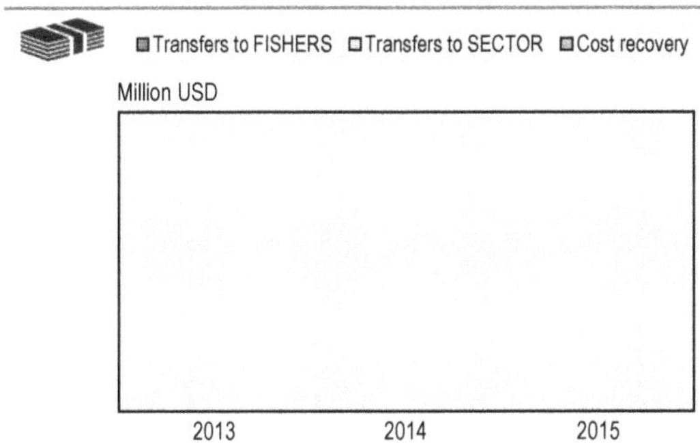

--- Number of stocks with established targets or thresholds recently subject to quantitative assessment

of which

--- Number of stocks found to be within acceptable conditions

Note: Reference year is 2014.

Source: OECD Fisheries Database. World's total values of production for catch and aquaculture were sourced from the *FAO Yearbook, Fishery and Aquaculture Statistics, 2015*. Both numbers have been corrected adding the value of seaweed production; aquaculture has also been adjusted to account for the value of Chinese production reported to the OECD.

Turkey

Developments in fisheries and aquaculture policy, 2015-2016

- Turkey is aligning its fisheries and aquaculture regulations with the recommendations of affiliated regional fisheries management organisations and the policy framework of the European Union. A special focus is placed on the development of environmentally friendly fisheries and aquaculture.
- The Government of Turkey financially supports the restructuring of its national fishing fleet aiming for balance between fishing capacity and available marine resources.
- The Government of Turkey is investing in improved control of its fisheries fleet. In 2016 a project introducing the Fishing Vessel Monitoring System was completed. Consequently, vessels landing 90% of the marine catch are subject to continuous monitoring. In addition, a system of electronic logbooks was implemented.

Fisheries and aquaculture statistics in Turkey

Value of production

in global context

USD 1 402 million	**0.4%**	**0.4%**
Total value of fisheries and aquaculture	*Share of global fisheries landed value*	*Share of global aquaculture production value*

Trends in production value and trade

Total production value in USD decreased at an annual average rate of 0.7% between 2011 and 2015

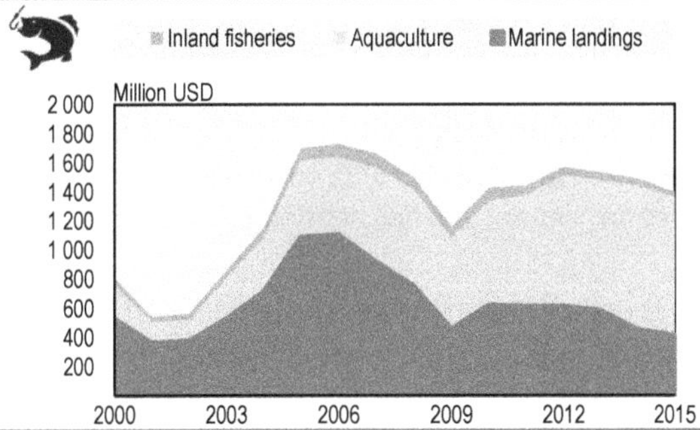

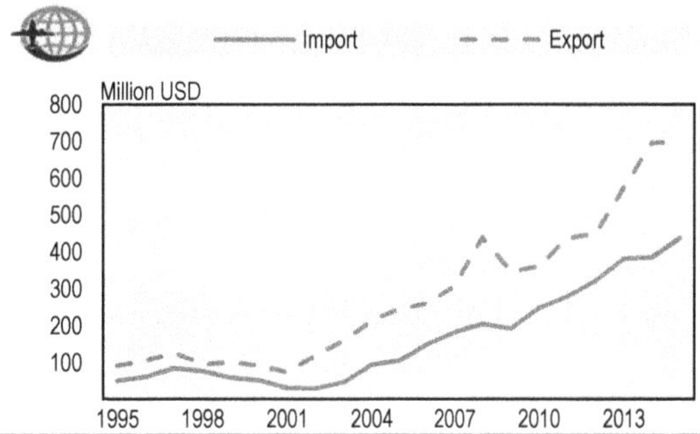

Employment and fleet

Employment in capture fisheries decreased by 2.4% between 2014 and 2015 to 35 821 persons

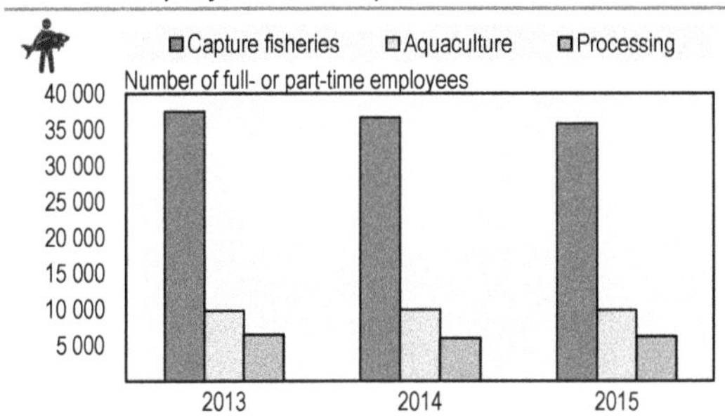

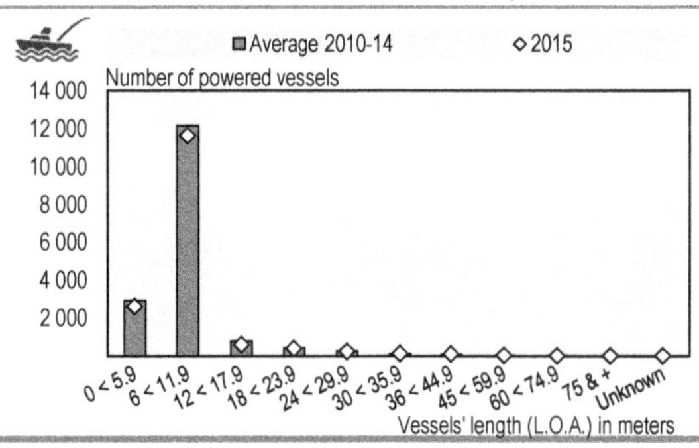

Government support and resource management

77% of total budgetary support was directed to management of resources in 2015

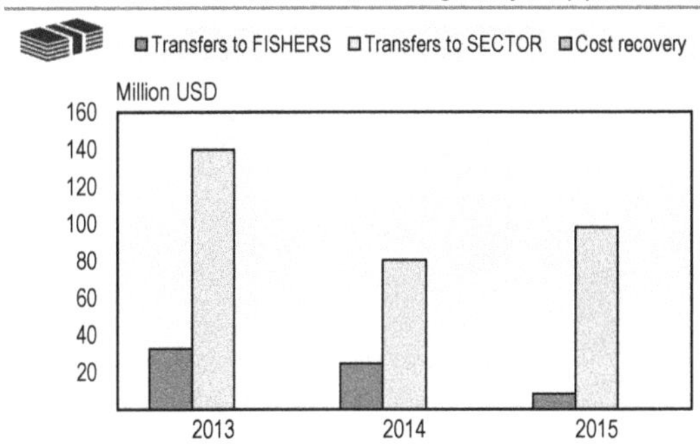

5 Number of Black Sea shared stocks, where riparien countries operate fisheries, with established targets or thresholds recently subject to quantitative assessment

of which

1 Number found to have acceptable status with respect to targets or thresholds

Note: Reference year is 2015, except for stock assessment data which may cover more than one year.

Source: OECD Fisheries Database. World's total values of production for catch and aquaculture were sourced from the *FAO Yearbook, Fishery and Aquaculture Statistics, 2015*. Both numbers have been corrected adding the value of seaweed production; aquaculture has also been adjusted to account for the value of Chinese production reported to the OECD. Fish stocks data were directly reported by the country.

United States

Developments in fisheries and aquaculture policy, 2015-2016

- In 2016, NOAA Fisheries finalised its Ecosystem-Based Fisheries Management (EBFM) Policy and Road Map that will facilitate more full integration of EBFM into its fisheries management decisions.
- In 2016, NOAA Fisheries established a rule implementing import provisions of the Marine Mammal Protection Act (MMPA) to reduce marine mammal bycatch.
- In May 2016, NOAA Fisheries released its Marine Aquaculture Strategic Plan to guide development of sustainable marine aquaculture and established the target of expanding US marine aquaculture production at least 50% by the year 2020.
- The 2016 Status of the Stocks Annual Report to Congress on the status of US fisheries reported that 91% of 313 stocks are not subject to overfishing. Two stocks were declared rebuilt in 2016.

Fisheries and aquaculture statistics in the United States

Value of production

in global context

USD 6 701 million — *Total value of fisheries and aquaculture*

4.6% — *Share of global fisheries landed value*

0.5% — *Share of global aquaculture production value*

Trends in production value and trade

Total production value in USD decreased at an annual average rate of 0.3% between 2011 and 2015

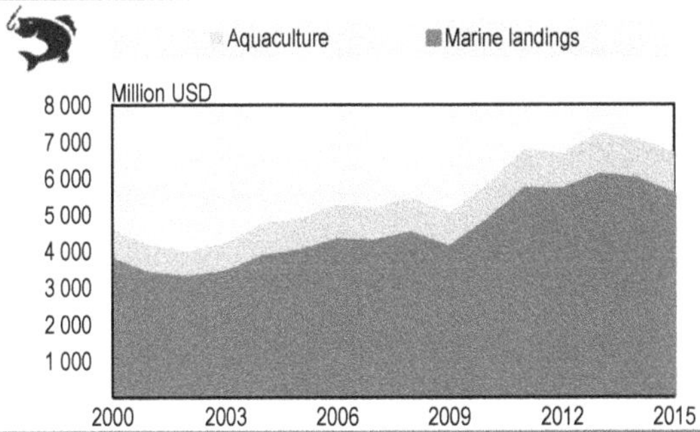

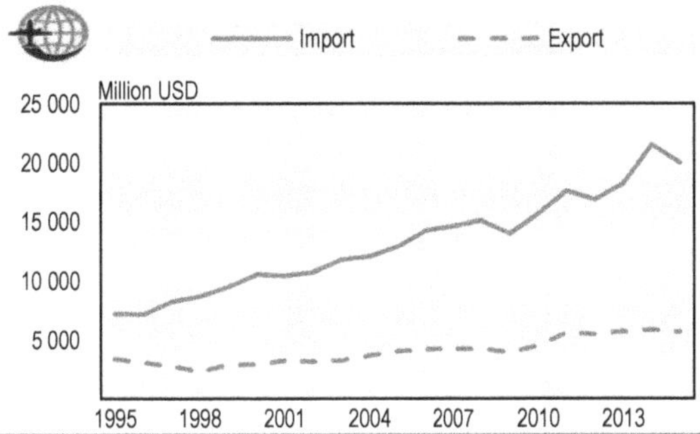

Employment and fleet

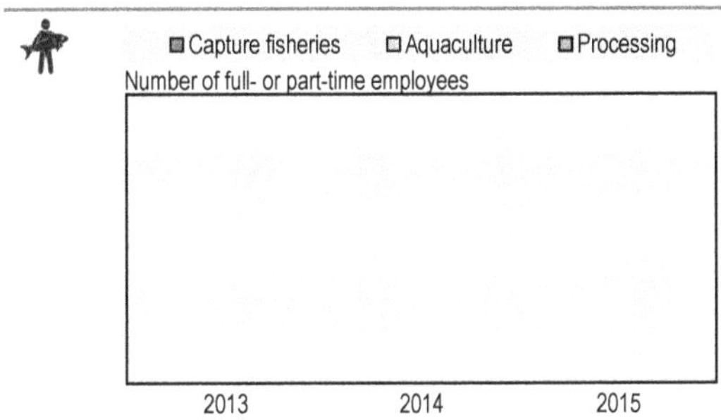

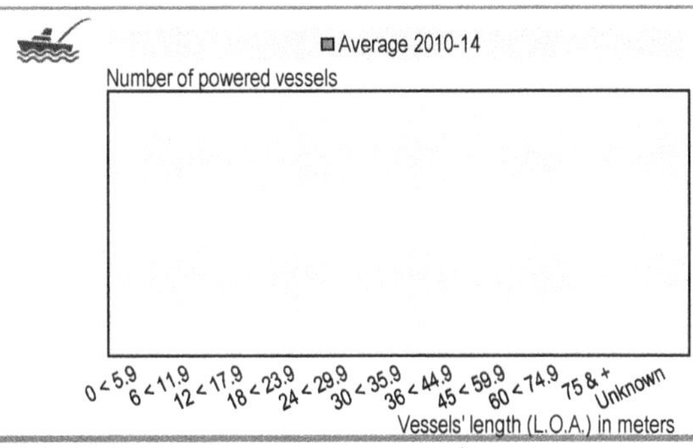

Government support and resource management

95% of total budgetary support was directed to management of resources in 2015

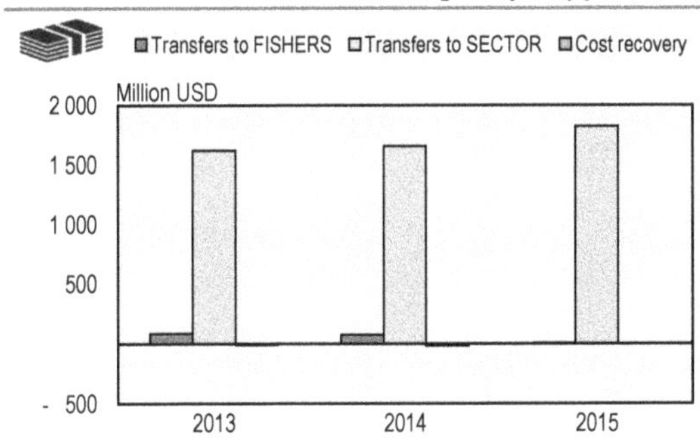

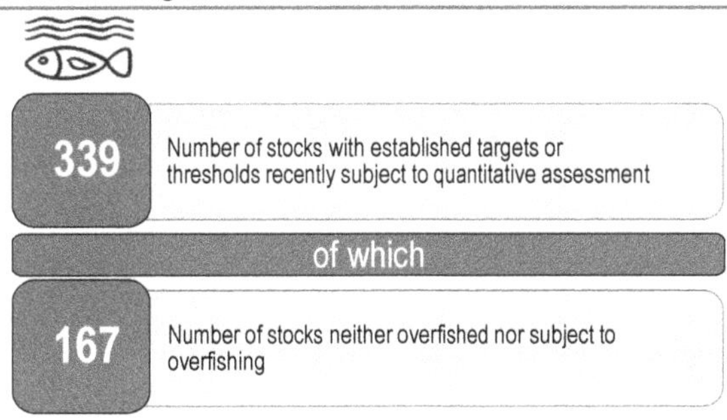

Note: Reference year is 2015, except for stock assessment data which may cover more than one year.
Source: OECD Fisheries Database. World's total values of production for catch and aquaculture were sourced from the *FAO Yearbook, Fishery and Aquaculture Statistics, 2015*. Both numbers have been corrected adding the value of seaweed production; aquaculture has also been adjusted to account for the value of Chinese production reported to the OECD. Fish stocks data were sourced from: the 4th Quarter 2016 Update of the FSSI and Non-FSSI Stock Status Table.

European Union

Developments in fisheries and aquaculture policy, 2015-16

- Recent developments in EU-wide fisheries related policies reflect the ambitious goal of sustainably exploiting fisheries by 2020, mainly through reform of the Common Fisheries Policy. Funding of EUR 5.7 billion (USD 7.6 billion) is allocated to achieve that goal under the European Maritime and Fisheries Fund (EMFF).
- Meeting the objectives of fisheries policies in the EU is challenging and although progress is steady, it is not uniformly distributed. Progress has been made in the North-East Atlantic region (including also the North Sea and the Baltic Sea), but moving towards maximum sustainable yield in the Mediterranean and the Black Sea remains challenging.
- Development of aquaculture is supported by the EU and is part of the reformed Common Fisheries Policy. In 2015, all Member States finalised Multiannual National Strategic Plans for the promotion of sustainable aquaculture in their countries. Aquaculture is an EMFF priority with EUR 1.2 billion (USD 1.6 billion) worth of support allocated to it.

ORGANISATION FOR ECONOMIC CO-OPERATION AND DEVELOPMENT

The OECD is a unique forum where governments work together to address the economic, social and environmental challenges of globalisation. The OECD is also at the forefront of efforts to understand and to help governments respond to new developments and concerns, such as corporate governance, the information economy and the challenges of an ageing population. The Organisation provides a setting where governments can compare policy experiences, seek answers to common problems, identify good practice and work to co-ordinate domestic and international policies.

The OECD member countries are: Australia, Austria, Belgium, Canada, Chile, the Czech Republic, Denmark, Estonia, Finland, France, Germany, Greece, Hungary, Iceland, Ireland, Israel, Italy, Japan, Korea, Latvia, Luxembourg, Mexico, the Netherlands, New Zealand, Norway, Poland, Portugal, the Slovak Republic, Slovenia, Spain, Sweden, Switzerland, Turkey, the United Kingdom and the United States. The European Union takes part in the work of the OECD.

OECD Publishing disseminates widely the results of the Organisation's statistics gathering and research on economic, social and environmental issues, as well as the conventions, guidelines and standards agreed by its members.

OECD PUBLISHING, 2, rue André-Pascal, 75775 PARIS CEDEX 16
(53 2016 01 1 P) ISBN 978-92-64-28225-4 – 2017

www.ingramcontent.com/pod-product-compliance
Ingram Content Group UK Ltd.
Pitfield, Milton Keynes, MK11 3LW, UK
UKHW051241300926
15471UKWH00015B/1448

9 789264 282254